Water Management and Water Loss

Water Management and Water Loss

Stuart Hamilton and Ronnie McKenzie

IWA PUBLISHING

Published by **IWA Publishing**
Alliance House
12 Caxton Street
London SW1H 0QS, UK
Telephone: +44 (0)20 7654 5500
Fax: +44 (0)20 7654 5555
Email: publications@iwap.co.uk
Web: www.iwapublishing.com

First published 2014

British Library Cataloguing in Publication Data
A CIP catalogue record for this book is available from the British Library

ISBN 9781780406350 (Paperback)
ISBN 9781780406367 (eBook)

Contents

Chapter 1
Introduction 1
1.1 The Technology Matrices 2
1.1.1 Main pipelines only – high pressure 2
1.1.2 Main pipelines only – low pressure 3
1.1.3 Domestic and mains fittings – high pressure 3
1.1.4 Domestic and mains fittings – low pressure 4

Chapter 2
Leak detection technologies 5
2.1 Method A: Gas Injection Method 5
2.2 Method B: Manual Listening Stick 5
2.3 Methods C and D: Leak Noise Correlation 6
2.4 Method C: Correlation Using Accelerometers 7
2.5 Method D: Correlation Using Hydrophones 8
2.6 Method E: In-Line Leak Detection Techniques 9
2.6.1 Tethered systems 10
2.6.2 Free swimming systems 11
2.7 Method F: Noise Loggers – Non-Correlating 12
2.8 Method G: Electronic Amplified Listening Devices 14
2.8.1 Step testing 15
2.8.2 Principles of step testing 15

Chapter 3
Principles of pressure management 19
3.1 Introduction 19
3.1.1 Basics of pressure management 19
3.1.2 Concepts of pressure management 23

3.1.3 Fixed outlet control 24
3.1.4 Time control 25
3.1.5 Flow control 26
3.1.6 Intelligent and closed-loop control 29
3.1.7 When to use advanced pressure control 30
3.1.8 Some large advanced pressure management installations 39
3.1.9 Khayelitsha: City of Cape Town – 2001 39
3.1.10 Sebokeng: Emfuleni Local Municipality – 2005 42
3.1.11 Mitchell's Plain: City of Cape Town – November 2008 46
3.1.12 Conclusions 49
References 50

Chapter 4
The repair or replace dilemma ***51***
4.1 Definitions of Repair and Replace 51
4.2 The Durability of a Network 51
4.3 Deterioration of a Distribution System 52
4.4 Factors Affecting the Risk of Failure of Water Distribution Pipes 52
4.5 Definition of Mains Failure 54
4.6 The Importance of Information 54
4.6.1 Problems with information collection 54
4.6.2 The 'Garbage In, Garbage Out' (GIGO) principal 54
4.6.3 Minimum requirements 54
4.6.4 Developing information systems 55
4.7 Distribution Performance Measures 55
4.7.1 The importance of measuring performance 55
4.7.2 Performance measures for distribution 55
4.7.3 CARE W 56
4.7.4 Cost information 56
4.7.5 Environmental factors 56
4.7.6 Configuring the data 57
4.8 A Mains Failure Database 57
4.8.1 Why mains failure information is needed 57
4.8.2 National failure databases 57
4.9 Starting from Scratch 58
4.9.1 Compiling a mains asset database 59
4.9.2 Capturing repair information 59
4.9.3 Collecting additional information 59
4.9.4 Organising the information 59
4.10 Analysing the Information 60
4.10.1 The simplest approach 60
4.10.2 Use of a GIS system 60
4.11 Dealing with High Burst Rates 62
4.11.1 How to decide if the failure rate is too high? 62
4.11.2 Pressure control to reduce failure rates 62

4.11.3 Increase of failure rates with increased pressure 63
4.11.4 Identifying mains replacement projects 63
4.12 Cost Benefit Studies 64
4.12.1 The need for cost benefit studies 64
4.12.2 Calculating direct costs 64
4.12.3 Potential savings 64
4.12.4 Comparing capital and operational costs 64
4.12.5 Indirect costs and triple bottom line approaches 64
4.13 Predicting Future Trends 65
4.13.1 Why predict the future? 65
4.13.2 Models available 65
4.14 Avoiding Leakage in New Systems 66
4.14.1 Choice of material 66
4.14.2 Contract requirements 66
4.15 Further Research 67
References 67

Chapter 5
Community awareness and education 69
5.1 Introduction 69
5.1.1 Importance of community awareness and education 69
5.1.2 Objectives of community awareness education 70
5.1.3 Methodologies for implementation 75
5.1.4 Lessons learnt 84
Bibliography 85

Chapter 6
Meter logging and recording 87
6.1 Introduction 87
6.1.1 General 87
6.1.2 Importance of logging 87
6.1.3 Interpretation of minimum night flows 88
6.1.4 Assessing the minimum night flow 88
6.1.5 Normal night use 90
6.1.6 Background leakage 91
6.1.7 Calculation of unexplained bursts 92
6.1.8 Using minimum night flow for leakage management 94
6.1.9 Live logging of flows at bulk consumer meters 102
References 107

Chapter 7
Large consumer meter consolidation and improvements: Planning, implementation and benefits 109
7.1 Introduction 109
7.1.1 Planning and policies 110

7.1.2 Implementation 113
7.1.3 Results 117
7.1.4 Conclusion 119
References 120

Chapter 8
Software tools to assist with water loss reduction in municipal systems ***121***
8.1 Introduction 121
8.1.1 Background 121
8.1.2 AQUALITE: Water balance software 123
8.1.3 SANFLOW: Background night flow analysis model 127
8.1.4 ECONOLEAK: Economics of leakage model 129
8.1.5 PRESMAC: Pressure management model 130
8.1.6 WDM scorecard 132
8.2 Latest Results from the NRW Assessments 137
Bibliography 141

Chapter 9
International case studies from around the world **143**

Chapter 9.1
Application in water loss reduction project in Bosnia and Herzegovina ***144***
Đevad Koldžo and Branko Vučijak
9.1.1 Introduction 144
9.1.2 Water Balance Evaluation Software – Calculeakator 144
9.1.3 Water Loss Reduction Project in DMA KUZICI (TUZLA) 146
9.1.4 Conclusion 149
References 150

Chapter 9.2
Large scale water loss management improvement programme in Zagreb (Croatia) water distribution network ***151***
Kristijan Iličić, Jurica Kovač, Vlado Herceg, Berislav Bohatka, Mario Medven and Ivan Grdenić
9.2.1 Introduction 151
9.2.2 DMAs and PMAs – Starting Points 152
9.2.3 Design of DMA/PMA 152
9.2.4 Implementation 154
9.2.5 Results 158
9.2.6 Conclusion 159
References 159

Chapter 9.3
Analysis of the water losses from the aspect of central water meters in objects of collective housing 160
Bajči Angela and Radovanov Milenko

9.3.1 Introduction 160
9.3.2 Description of the Problem 161
9.3.2.1 Example problem 1 – leakage in buildings 162
9.3.2.2 Example problem 2 – leakage in buildings 162
9.3.2.3 Example problem 3 – leakage in buildings 163
9.3.2.4 Example problem 4 – water meter diameter 164
9.3.3 Analysis 164
9.3.3.1 User coverage 164
9.3.3.2 Water consumption and flat-rate collection 165
9.3.3.3 Analysis – losses due to flat-rate collection of payment 167
9.3.3.4 Results of the analysis 168
9.3.4 Conclusion 168
References 169

Chapter 9.4
Water loss management in Novi Sad water utility, Serbia 170
Zlatko Arvaji and Ikonija Karadžić

9.4.1 Introduction 170
9.4.2 Pressure Management 170
9.4.3 District Metered Areas and Basic Real Loss Performance Indicators 171
9.4.3.1 Micro metered areas 171
9.4.4 Active Leakage Control and the Pilot DMAs 172
9.4.5 Water Price Policy and Water Meter Inaccuracies 173
9.4.6 Strategy 2014 174
9.4.7 Conclusion 174
References 175

Chapter 9.5
Continuous acoustic monitoring – evolution of an innovation in the USA 176
David Hughes

Chapter 9.6
Establishing the first validated dataset of North American water utility water audit data 182
A. Chastain-Howley, G. Kunkel and W. Jernigan

9.6.1 Methodology 183
9.6.2 Results 184

9.6.2.1 Analysis of DV scores in dataset 189
9.6.2.2 Trend review 190
9.6.2.3 System size 190
9.6.2.4 Climate 190
9.6.2.5 Connection density 191
9.6.2.6 Cost data 191
9.6.2.7 Improving Data Validity 191
9.6.3 Conclusions 191
References 192

Chapter 1

Introduction

Water losses from municipal reticulation systems are becoming a serious problem throughout the world and particularly in developing countries where lack of funds often results in poor maintenance of the water infrastructure. In some areas, the water losses are now estimated to be higher than the actual legitimate water use and the situation is gradually deteriorating to such an extent that intermittent supply is now the norm rather than the exception in some parts of the world.

The International Water Association has identified water loss from municipal systems as one of the most important issues facing a world where the ever increasing population is placing additional strain on systems which are already failing to meet the current demands. In response to these problems, the Water Losses Specialist Group was formed and has emerged as one of the most active and significant groups within the IWA. Its key function is to share the knowledge from specialists throughout the world to assist water suppliers to manage and reduce their water losses. Through the efforts of the group, many new and innovative techniques have been developed to measure, monitor and analyse water losses from municipal water supply systems. A standard water balance has been developed which is now widely accepted worldwide and enables water suppliers to quantify the magnitude of their water losses in a standard and pragmatic approach after which they can select the most appropriate interventions for a specific area. No "one-size fits all" solution is appropriate when trying to deal with water losses and it is important to understand that each area has its own set of problems, some of which may be common to other areas and some of which may be unique to a specific region. It is therefore necessary to identify the key problems and to implement the appropriate interventions.

This book provides a basic overview of the key issues frequently experienced by water supply managers in both developing and developed countries. It is designed to assist water supply managers to understand the problems experienced in their own systems and provides advice on how the problems can be addressed through examples and practical case studies. The book is non-academic and is aimed at providing information and advice in a practical and easy to understand manner. The book is basically a collection of case studies and practical experience from numerous water loss specialists who have worked around the world and have experienced both 'best practice' and the opposite extreme. It is hoped that by providing such examples and case studies, water supply managers and those employed to assist them, can benefit from both the successes and failures discussed in the book.

1.1 THE TECHNOLOGY MATRICES*

The choice of a particular leak detection/location technique and technology depends on the operating conditions and construction material of the pipeline in question. To assist in making this determination, four different matrices have been developed.

(1) Mains fittings only – high pressure
  - For leakage detection on mains fittings only (no house connections) with pressures greater than 10 m head or 15 psi. Fittings are at a minimum distance of 200 m apart and maximum 500 m

(2) Mains fittings only – low pressure
  - For leakage detection on mains fittings only (no house connections) with pressures less than 10 m head or 15 psi. Fittings are at a minimum distance of 200 m apart and maximum 500 m

(3) Domestic and mains fittings – high pressure
  - For leakage detection on all property and mains fittings with pressures greater than 10 m head or 15 psi. Fittings are at a minimum distance of 10 m apart and maximum 50 m

(4) Domestic and mains fittings – low pressure
  - For leakage detection on all property and mains fittings with pressures less than 10 m head or 15 psi. Fittings are at a minimum distance of 10 m apart and maximum 50 m

The matrices consider the following pipeline materials:

- Metallic
  - Includes steel, ductile iron and other ferrous materials
- Concrete
  - Includes reinforced concrete, pre-stressed concrete pipe (PCP)
- Asbestos cement
- Glass-reinforced plastic (GRP)
- Polyvinyl chloride (PVC)
- Polyethylene
  - MDPE medium density poly ethylene
  - HDPE high density poly ethylene

The technologies available are discussed in more detail later in this document. The equipment has been placed in the selected categories where it is reliably successful. The equipment may sometimes be successful in other categories but not reliably so.

Note that new equipment is continuously being developed: these matrices only take into account equipment that was available during the preparation of the matrices (up to December 2012).

1.1.1 Main pipelines only – high pressure

This matrix is for leakage detection on mains fittings only (no house connections) with pressures greater than 10 m head or 15 psi. Fittings are at a minimum distance of 200 m apart and maximum 500 m.

*This section is from Chapter 2 of *Leak Detection: Technology and Implementation* by Stuart Hamilton and Bambos Charalambous, Published in 2013 by IWA Publishing.

Diameter mm	75	100	150	200	250	300	350	400	450	500	600	700	800	900	1000+
inches	3	4	6	8	10	12	14	16	18	20	24	28	32	36	40+
Material															
Metallic all	A,B, C,D, F,G	A,B, C,D, F,G	A,B, C,D, F,G	A,B, C,D, F,G	A,B, C,D, F,G	A,C, D,E, F,G	A,C, D,E, F,G	A,C, D,E	C,D,E	C,D,E	D,E	D,E	E	E	E
Concrete all	A,C,D	A,C,D	A,C,D	A,C,D	A,D	A,D,E	A,D,E	A,D,E	E	E	E	E	E	E	E
Asbestos Cement	A,C,D	A,C,D	A,C,D	A,C,D	A,D	A,D,E	A,D,E	A,D,E	E	E	E	E	E	E	E
GRP	A,D	A,D	A,D	A,D	A,D	A,D,E	A,D,E	A,D,E	E	E	E	E	E	E	E
PVC	A,D	A,D	A,D	A,D	A,D	A,D,E	A,D,E	A,D,E	E	E	E	E	E	E	E
Polyethylene all	A,D	A,D	A,D	A,D	A,D	A,D,E	A,D,E	A,D, E	E	E	E	E	E	E	E

Method A Gas Injection
Method B Traditional Techniques with Manual Listening Stick
Method C Non-Intrusive Acoustic Techniques that is Standard Correlator, Correlating Noise Loggers (Accelerometers)
Method D Intrusive Acoustic Techniques that is Standard Correlator or Correlating Noise Loggers (Hydrophones)
Method E Inline Inspection Techniques (Tethered and Free-swimming)
Method F Noise Loggers (Non-Correlating), Non-Intrusive Magnetic Connection
Method G Electronic Amplified Listening Ground Microphone

1.1.2 Main pipelines only – low pressure

This matrix is for leakage detection on mains fittings only (no house connections) with pressures less than 10 m head or 15 psi. Fittings are at a minimum distance of 200 m apart and maximum 500 m.

Diameter mm	75	100	150	200	250	300	350	400	450	500	600	700	800	900	1000+
inches	3	4	6	8	10	12	14	16	18	20	24	28	32	36	40+
Material															
Metallic all	A,D	A,D	A,D	A, D	A,D	A,D,E	A,D,E	A,D,E	E	E	E	E	E	E	E
Concrete all	A,D	A,D	A,D	A, D	A,D	A,D,E	A,D,E	A,D,E	E	E	E	E	E	E	E
Asbestos Cement	A,D	A,D	A,D	A,D	A,D	A,D,E	A,D,E	A,D,E	E	E	E	E	E	E	E
GRP	A,D	A,D	A,D	A,D	A,D	A,D,E	A,D,E	A,D,E	E	E	E	E	E	E	E
PVC	A,D	A,D	A,D	A,D	A,D	A,D,E	A,D,E	A,D,E	E	E	E	E	E	E	E
Polyethylene all	A,D	A,D	A,D	A,D	A,D	A,D,E	A,D,E	A,D,E	E	E	E	E	E	E	E

Method A Gas Injection
Method B Traditional Techniques with Manual Listening Stick
Method C Non-Intrusive Acoustic Techniques, that is, Standard Correlator, Correlating Noise Loggers (Accelerometers)
Method D Intrusive Acoustic Techniques, that is, Standard Correlator or Correlating Noise Loggers (Hydrophones)
Method E Inline Inspection Techniques (Tethered and Free-swimming)
Method F Noise Loggers (Non-Correlating), Non-Intrusive Magnetic Connection
Method G Electronic Amplified Listening Ground Microphone

1.1.3 Domestic and mains fittings – high pressure

This matrix is for leakage detection on all property and mains fittings with pressures greater than 10 m head or 15 psi. Fittings are at a minimum distance of 10 m apart and maximum 50 m.

Diameter	mm	75	100	150	200	250	300	350	400	450	500	600	700	800	900	1000+
	inches	3	4	6	8	10	12	14	16	18	20	24	28	32	36	40+
Material																
Metallic all		A,B,C, D,F,G	A,B,C, D,F,G	A,B,C, D,F,G	A,B,C, D,F,G	A,B,C, D,F,G	A,C,D, E,F,G	A,C,D, E,F,G	A,C,D, E,F,G	C,D,E, F,G	C,D,E, F,G	C,D, E	C,D, E	D,E	D,E	D,E
Concrete all		A,C, D,F,G	A,C, D,F,G	A,C, D,F,G	A,D	A,D	A,D,E	A,D,E	A,D,E	E	E	E	E	E	E	E
Asbestos Cement		A,C, D,F,G	A,C, D,F,G	A,C, D,F,G	A,C, D	A,C, D	A,D, E	A,D, E	A,D, E	E	E	E	E	E	E	E
GRP		A,C, D,F,G	A,C, D,F,G	A,C, D,F,G	A,C, D	A,C, D	A,D, E	A,D, E	A,D, E	E	E	E	E	E	E	E
PVC		A,C, D,F,G	A,C, D,F,G	A,C, D,F,G	A,D	A,D	A,D, E	A,D, E	A,D, E	E	E	E	E	E	E	E
Polyethylene all		A,C, D,F,G	A,C, D,F,G	A,C, D,F,G	A,D	A,D	A,D, E	A,D, E	A,D, E	E	E	E	E	E	E	E

Method A Gas Injection
Method B Traditional Techniques with Manual Listening Stick
Method C Non-Intrusive Acoustic Techniques that is, Standard Correlator, Correlating Noise Loggers (Accelerometers)
Method D Intrusive Acoustic Techniques that is, Standard Correlator or Correlating Noise Loggers (Hydrophones)
Method E Inline Inspection Techniques (Tethered and Free-swimming)
Method F Noise Loggers (Non-Correlating), Non-Intrusive Magnetic Connection
Method G Electronic Amplified Listening Ground Microphone

1.1.4 Domestic and mains fittings – low pressure

This matrix is for leakage detection on all property and mains fittings with pressures less than 10 m head or 15 psi. Fittings are at a minimum distance of 10 m apart and maximum 50 m.

Diameter	mm	75	100	150	200	250	300	350	400	450	500	600	700	800	900	1000+
	inches	3	4	6	8	10	12	14	16	18	20	24	28	32	36	40+
Material																
Metallic all		A,C, D,F	A,C, D,F	A,C, D,F	A,C, D,F	A,D	A,D,E	A,D,E	A,D,E	E	E	E	E	E	E	E
Concrete all		A,D	A,D	A,D	A,D	A,D	A,D,E	A,D,E	A,D,E	E	E	E	E	E	E	E
Asbestos Cement		A,D	A,D	A,D	A,D	A,D	A,D,E	A,D,E	A,D,E	E	E	E	E	E	E	E
GRP		A,D	A,D	A,D	A,D	A,D	A,D,E	A,D,E	A,D,E	E	E	E	E	E	E	E
PVC		A,D	A,D	A,D	A,D	A,D	A,D,E	A,D,E	A,D,E	E	E	E	E	E	E	E
Polyethylene all		A,D	A,D	A,D	A,D	A,D	A,D,E	A,D,E	A,D,E	E	E	E	E	E	E	E

Method A Gas Injection
Method B Traditional Techniques with Manual Listening Stick
Method C Non-Intrusive Acoustic Techniques that is, Standard Correlator, Correlating Noise Loggers (Accelerometers)
Method D Intrusive Acoustic Techniques that is, Standard Correlator or Correlating Noise Loggers (Hydrophones)
Method E Inline Inspection Techniques (Tethered and Free-swimming)
Method F Noise Loggers (Non-Correlating), Non-Intrusive Magnetic Connection
Method G Electronic Amplified Listening Ground Microphone

Chapter 2

Leak detection technologies*

There are a vast number of techniques to detect where leakage is occurring in the network. Location accuracy depends on many factors, and the subsequent portions of this document provide further detail. Some techniques are able to approximate or localize the position of a leak while others can find exact locations. Often a tool-box approach is used, where multiple technologies are deployed.

2.1 METHOD A: GAS INJECTION METHOD

This method uses a gas detector to find the presence of a tracer gas that has been injected into a pipeline. While helium can be used, the most common tracer gas is hydrogen due to its lower cost and high performance.

Hydrogen is the lightest gas and has the lowest viscosity. This makes it easy to fill, evacuate and dissipate. Typically diluted 5% in nitrogen, the gas can be injected into buried and ducted cables, pipelines and also small diameter in-house heating pipes.

The gas injection method can be used to detect leaks in all pipe materials from 75 mm to 1000 mm in diameter. It can be used on pipes of greater diameter but for obvious reasons a considerable amount of gas would be required. The pipeline can be empty of water or full, however with the pipeline full of water, less gas is required to be used to find the leak.

To accurately locate the leaking gas which comes to the surface after leaving the leak in the pipe, the direction of the water flow must be known and the gas should be kept within the pipeline in which the leak is suspected. This requires the closure of any branches/off-takes which may cause the gas to be diluted or transferred away from the pipeline in question. The mixing of the gas with water does not affect the water quality. This methodology can be used in all types of sealed tubes including cables and pipelines. The material has no effect on the gas injected.

2.2 METHOD B: MANUAL LISTENING STICK

The stethoscope or listening stick has an earpiece and is used to listen to leaks in fittings and to pinpoint the location of a leak. It is a widely used piece of equipment for many water utilities. The

*This chapter is modified from Chapter 4 of *Leak Detection: Technology and Implementation* by Stuart Hamilton and Bambos Charalambous, Published in 2013 by IWA Publishing.

material of the listening stick can be metal, wooden or plastic. This technique is dependent on the ability of the engineer to hear the leak and uses no electronic equipment to enhance the sound.

This technique is best suited for use on metallic pipelines between 75 mm and 250 mm and with pressures above 10 m (15 psi). The material or pipe size does not prevent the listening stick from being able to pinpoint the leak from the surface, but what does affect this is the type of leak, ground backfill material, pressure of the water leaving the pipe, background noise and the ability of the engineer (Figure 2.1).

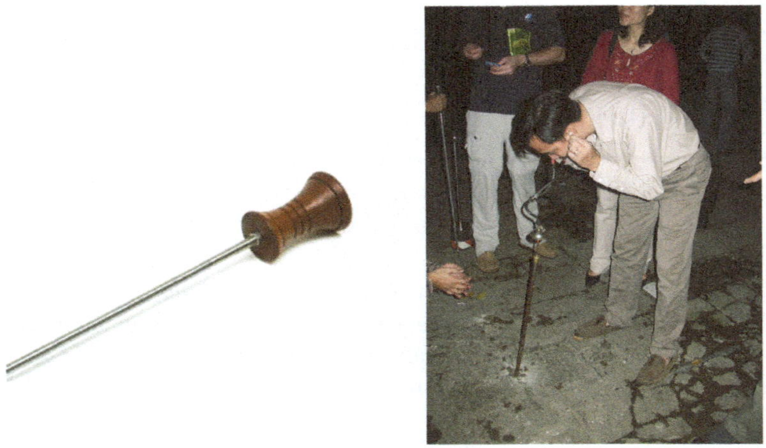

Figure 2.1 Manual listening sticks.

2.3 METHODS C AND D: LEAK NOISE CORRELATION

Leak noise correlation works by comparing the noise detected at two different points in the pipeline. Assuming consistent pipe material and diameter, the noise travels from the leak in both directions at a constant velocity, so that if the leak is equidistant between two sensors then these sensors will detect the noise at the same time. Conversely, if the leak is not equidistant, then the sensors will detect the same noise at different times – this difference in arrival times is measured by the correlation process.

The following diagram illustrates this principle (Figure 2.2):

The sensors are located on valves A and B (convenient access points for underground pipes), and as shown, the leak position is closer to A.

By the time an instance of noise from the leak has reached A, the same noise heading towards B has only travelled as far as point X. The distance from X to B causes a time delay (t) before the noise arrives at B. The correlation processing detects the delay (t) between the arrival of the noise at A and its arrival at B. If the velocity of sound is V and the distance between the loggers is D, then as the distance from X to B = $V * t$.

Then $D = (2 * L) + (V * t)$.

This equation may be rearranged to give L, the distance from the nearer logger to the leak site:

$$L = \frac{D - (V * t)}{2}$$

Correlation measures the time delay (*t*). The distance between the sensors must be determined by accurate measurement.

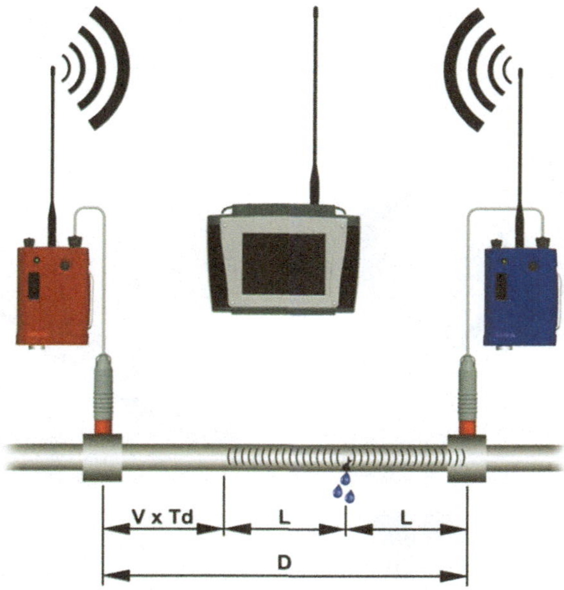

Figure 2.2 Principle of correlation (*Source*: Halma Water Management).

The sound velocity depends upon pipe material, pipe diameter and, to a lesser extent, on surrounding soil. Often, theoretical values of sound velocity are used and this is fine for a first approximation of the leak position. However, the velocity will vary due to many factors, and significantly so if a repair section of a different pipe material exists. Sound velocity must therefore be measured or, alternatively, multiple correlations carried out.

With all correlation techniques, practitioners should be aware that any noise source can result in a correlation peak and all results should thus be treated as 'points of interest' until confirmed. Confirmation is usually done using a ground microphone.

It is important to note that the capability of correlators is dependent on the pressure and level of background noise within the network. Furthermore, correlation can become impossible, because it requires two monitoring points, one on each side of the leak and the attenuation often causes leak signals to disappear at one or both points.

Leak noise correlation requires a noise signal. There are two types of noise transducer in normal use: accelerometers and hydrophones. These have significant differences in their deployment and uses, and have hence been identified as two different methods in the sections below.

2.4 METHOD C: CORRELATION USING ACCELEROMETERS

To perform correlation using accelerometers, two sensors are deployed on pipe fittings, so no access to the water inside the pipe is required. The sensors are then positioned on either side of the suspect leak position. Accelerometers respond to acceleration and so tend to be more responsive to higher frequencies

Accelerometers are most effective on metallic pipes and tend to be less effective with non-metallic pipes. This is due to both the rapid attenuation of high frequency signals in many non-metallic pipes and the impedance mismatch between the pipe material and the metal fittings normally used to site accelerometers. However their ease of deployment and low-cost make them an attractive choice when trying to localize leaks (Figure 2.3).

Figure 2.3 Example of accelerometer (*Source*: Halma Water Management (left) and primayer (right and bottom)).

2.5 METHOD D: CORRELATION USING HYDROPHONES

Leak location on plastic pipes and in large diameter pipes is difficult because they become more elastic as the pipe material gets softer, or the ratio of pipe wall thickness to overall diameter increases. This results in more rapid absorption of leak energy into the pipe material and surrounding soil, with resultant attenuation as distance from the leak increases.

Furthermore, higher frequency leak noise energy is absorbed much more rapidly than low frequency energy.

Accelerometers placed on the outside of the pipe detect the energy lost into the pipe wall. The water-borne wave is detected directly by hydrophones placed into the water at convenient fittings such as fire hydrants. Combined with advanced filtering technology, particularly at very low acoustic frequencies, hydrophones are beneficial in enabling leak location on plastic pipes and on larger diameter mains (Figure 2.4).

Figure 2.4 Example of hydrophone (*source*: primayer (top) and seba KMT (bottom)).

Hydrophone-based acoustic correlation techniques are beneficial in locating leaks in difficult situations. These situations are likely to occur in the following:

- Large diameter or trunk mains
- Plastic pipes
- Pipes with large distances between available pipe access points
- When high acoustic background noise exists, often due to traffic, and so on.

2.6 METHOD E: IN-LINE LEAK DETECTION TECHNIQUES

Specifically designed for large diameter pipes, in-line systems are able to discriminate between multiple leaks in a single length of pipeline. Pipelines can be inspected while under pressure and in service, and leaks are accurately located. Equipment is suitable for potable water applications, and no disruption to customers is needed.

There are two types of in-line systems; tethered and free swimming. Both have their advantages and disadvantages, but in both cases a sensor passes directly beside leaks, meaning that neither the pipe material nor the type of leak is relevant. The proximity of the sensor to any leak also results in very sensitive systems and the detection of small leaks.

2.6.1 Tethered systems

Tethered systems operate by deploying a hydrophone into the pipeline to be inspected. The hydrophone is connected to a signal processing and display unit via an umbilical cable. The sensor travels along inside the pipe pulled by the flow of water acting on a drogue (parachute) attached to the front of the sensor. As the sensor passes any leak in the pipeline it will detect the noise being generated by the water escaping through the leak. At this point the operator is able to stop deployment of the sensor (by stopping deployment of the umbilical) and then position the sensor at the leak position by withdrawing or deploying the umbilical as necessary.

Once the sensor is sited at the leak, the position of the sensor can be determined using a locating system mounted in the sensor head. A second operator can track the position of the sensor head during deployment using this locating device giving an accurate indication of the sensor location and pipe track. Having pinpointed the position of the sensor, the exact location of any leak can be marked on the ground over the pipe (Figure 2.5).

Figure 2.5 Tethered system gives access through 48 mm or above connection and only goes with the flow (*Source*: Pure Technologies and JD7).

Tethered systems are best suited to work in relatively straight pipelines where deployments of up to 2 km from a single insertion point are possible; other constraints may restrict the length that can be surveyed from each insertion point. Careful planning of the work will maximize the distance that can be surveyed.

Tethered systems are deployed through a tapping made into the pipe. Typically these insertion points may be air-valve connections, insertion flowmeter tappings or specially installed points.

Tethered systems that can access the main through fire hydrants are also available. Alternative sensors, including video and ultrasonic pipe-wall inspection, are also available (Figure 2.6).

Figure 2.6 Push tethered system which can be operated with or against the water flow with access through tapping point or fire hydrant (*Source*: JD7).

2.6.2 Free swimming systems

Free-swimming data acquisition systems for detecting leaks in fluid pipelines are also available. These systems are inserted into a live pipeline and pushed along the pipeline by the water flow. At the end of the inspection, a net or similar capture device is used to catch and extract the system from the pipeline or they are discharged into an open catchment area such as a reservoir and recovered from there; an acoustic recording is made during the entire inspection.

Free swimming devices are also available that during the free flowing inspection of the system have the capability of capturing CCTV footage and acoustic noise during the same survey.

Leaks are identified during data analysis following the removal of the system. These systems consist of the following components:

(a) Some of the free-swimming sensors contain all or just some of the following: an internal CCTV camera, acoustic sensor, tracking sensors, acoustic transponder, data processor, memory device and batteries.
(b) Above-ground tracking devices (which are used to track the progress of the sensor through the pipe).
(c) Insertion equipment/launch tube.
(d) Retrieval equipment.

The maximum length of pipeline that can be surveyed is determined by the flow rate in the line. For instance, with a flow rate of 1 m/s and a maximum operating life of 12 hours, the system can survey 43 km from a single insertion point; some of the available technology can record for 24 hrs, hence this distance can be doubled. All systems available today can traverse around tight bends and through inline valves (Figures 2.7 and 2.8).

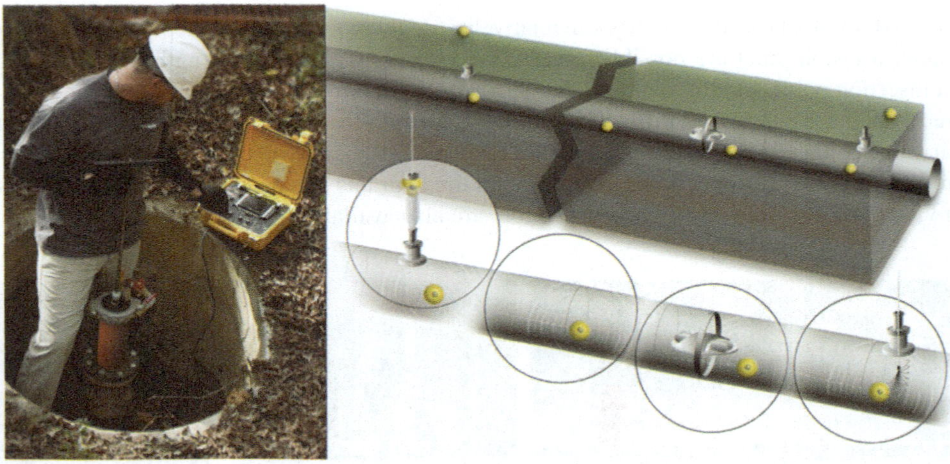

Figure 2.7 Free-swimming sensor deployment schematic (*Source*: Pure Technologies).

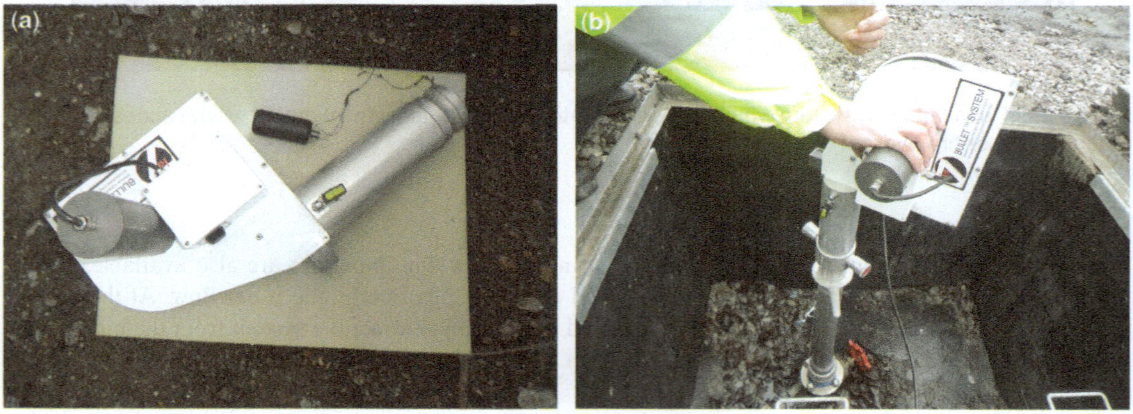

Figure 2.8 (a) Free flowing device combined CCTV, acoustic head, tracking device and launch tube; (b) free flowing device launched into main (*Source*: JD7 Ltd).

The on-board instrumentation allows the velocity of the system at all points along the survey route to be calculated during post-processing. This, combined with the use of above ground tracking devices, allows for the accurate location of any leaks (Figure 2.13).

2.7 METHOD F: NOISE LOGGERS – NON-CORRELATING

Previous sections show how noise is created by a leak and propagates through a pipe. Leak noise loggers are designed to 'pick up' this leak noise by being placed on available fittings, usually with a magnetic coupling (Figure 2.9).

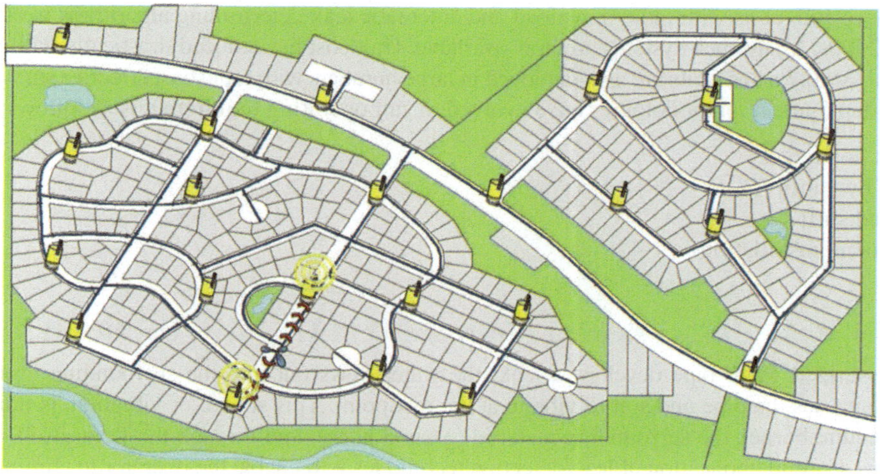

Figure 2.9 Noise logger schematic (*Source*: Halma Water Management).

The leak is identified by each logger unit individually based on the noise signature of a leak being consistent and loud against the background noise. Typically, measures of noise and consistency together with a graphical representation are supplied to the operator. The leak position is 'localized' to being between two loggers for follow-up pinpointing. The objective is to survey large areas at low cost to maximize efficiency in an active leak detection strategy (Figure 2.10).

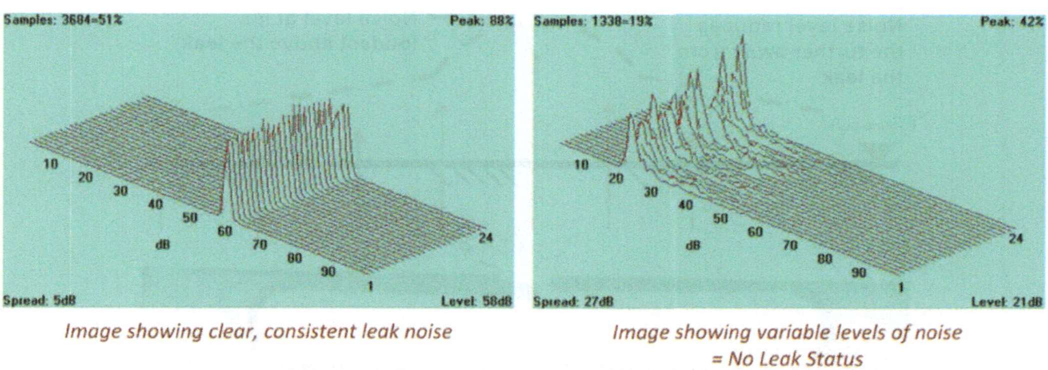

Figure 2.10 Typical leak/no leak display (*Source*: Halma Water Management).

Loggers are usually programmed to log during the middle of the night where interfering noise (traffic, legitimate water usage, etc.) is at a minimum and leaks can be most easily 'heard'. Exact sampling regimes vary across the available systems.

Early systems were manually programmed and downloaded with the operator determining whether or not a leak was present from data supplied. As mass deployment evolved, the need to further automate and ease the process became apparent with the objective being to survey large areas quickly and 'automatically' at low cost.

To the above end, units with radio download and automatic leak determining algorithms were introduced in the early 2000s. Many such systems are now available. Unit cost has reduced drastically with volume and technological evolution and units are now deployed in large numbers with rapid cost effective surveys possible.

Multiple deployment methodologies have evolved to suit operating requirements as follows:

- Direct download
- Drive-by patrol
- Lift and shift
- Permanent installation.

2.8 METHOD G: ELECTRONIC AMPLIFIED LISTENING DEVICES

When a pressurized water pipe develops a leak, the water flows out into the surrounding ground at high speed, which causes the pipe and soil to vibrate at the exit point. This sound, or vibration, is transmitted by the pipe (structure borne), the surrounding material (ground borne) and through the water itself over a range of frequencies. Careful application of leak detection techniques will enable the operator to eliminate detected noises generated by poor pipeline design or consumer usage and to identify leakage due to pipe system damage.

In addition to being transmitted along the pipeline (both through the water and the pipe wall) the leak noise is transmitted into the ground around the pipe. The noise travels much better through 'hard' materials so that the noise travels much further along metallic pipes than asbestos cement pipes, which themselves are better than plastic pipes. Ground material generally provides a poorer travel path than the pipeline itself. However, usually some noise is transmitted the short distance to the surface. Soft sandy ground provides a worse travel path than well compacted ground with a hard paved surface covering (Figure 2.11).

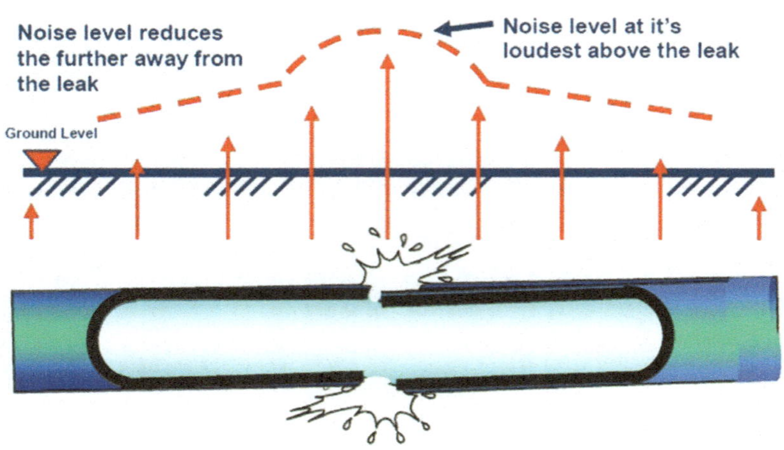

Figure 2.11 Schematic showing propagation of leak noise through ground (*Source*: Halma Water Management).

Factors producing good quality leak noise include high water pressure, hard backfill, a small rupture, clean pipes, metallic pipes and small diameter pipes. By contrast, factors producing poor quality leak noise include low water pressure, soft backfill, split mains, encrusted pipes, soft/lined pipes and large diameter pipes.

Since 'leak detection' began, operators have been 'listening' for this leak noise using mechanical devices. Traditional listening sticks for detecting water leaks rely on only one of the user's senses – hearing – the experience and skill of the operator is paramount and, at best, users are only ever able to detect leaks that produce loud noises.

However, it must be noted that not all leaks produce a noise audible to the human ear. Contrary to common perception, it is not always the largest leaks which are the loudest; often a large split in a water pipe will produce a less clear noise than a small hole. This can be particularly true in PVC, PE and MDPE pipe materials. For this reason, amplifying the noise with an electroacoustic microphone is becoming increasingly important to finding leaks, particularly in networks where these materials are increasingly used.

Modern electronics therefore provide the benefits of advanced sensor technology amplification and filtering to undertake this operation more effectively.

2.8.1 Step testing

Step testing is an effective, flow-based method of localizing water loss within a zoned distribution system. It is particularly suited to identifying areas of high leakage and to use on plastic pipe materials, where leak noise is absorbed and conventional acoustic methods are less effective.

To perform a step test, the inflow into a zone must be monitored. This can be achieved by deploying a data logger upon the inlet water meter to automatically transmit flow data to the operative in the network. Alternatively an additional operator can be left upon the inlet meter to manually record flow and network activity.

Once a method of monitoring has been established, the valves are closed to cut off sections of the zone known as 'steps'. This demonstrates how much water is consumed in each step. Each step has an estimated customer consumption which is compared to the drop in flow at the inlet meter. If the difference between the actual drop in demand and estimated consumption is significant, this provides an indication to the operative that leakage is contained within that step.

There are several different variations on the approach depending on the technology available, whether it is important to maintain supply and the configuration of the network.

2.8.2 Principles of step testing

- Each zone will require a plan that identifies which pipe lengths are to be used, valves to shut and in which order valves are to be shut. This should be used for all subsequent step tests, providing the zone does not change. By keeping the plan consistent, the operators can provide further judgement based upon experience.
- When designing a step test plan it is important to have an optimum number of 'steps'. This will largely depend upon the size of the zone. Too few steps may not achieve the desired reduction in leakage detection time and costs. Too many 'steps' can also be time consuming and the rate of leakage may be too insignificant for the flow meter to register.
- Another consideration when designing 'steps' is to calculate an estimation of customer consumption so that the operative has an expectation of a typical flow rate into a 'step'.
- Step tests should be carried out when demand is at its lowest. This tends to be at night time between the hours of 01:00–04:00. This helps contribute to a more accurate step test as fluctuations in demand are minimized.
- Before a step test is implemented, all valves required must be located on site. Once located then the integrity of the valve must be tested. This will include ensuring the valve is accessible and operable.

A Zero Pressure Test (ZPT) can conclude if the valve can be closed completely without passing any water, this helps contribute to further reassurance of an accurate step test.

- There are two options in providing essential flow data to the operator closing valves in the network:
  (1) Another operator upon the inlet meter with a telephone/radio communication to the operator in the network operating valves
  (2) Radio/GPRS data logger to transmit flow data to a suitable receiving device for the operator in the network to see live flow data (Figure 2.12).
- There are three types of valves when operating a step test:
  (1) Valves that are permanently shut to create a zone. These can sometimes be called boundary valves or zone valves.
  (2) Valves that need to be shut before the start of the test in order to create 'steps' that can be closed off during the test with a single valve, as sometimes it is not possible to shut off a section by using only one valve closure. These valves are only shut for the duration of the test and re-opened once the test is completed. They are sometimes known as circulating valves.
  (3) The final type of valve is one that isolates a step from the zone. They are numbered according to the order that they need to be shut in during the test. Step 1 is typically the 'step' that is the furthest away from the meter and the last step closure is the one nearest the meter.
- It is important to allow a settling time (approximately 15 minutes) between each 'step' closure so that a stable and realistic flow rate can be obtained.
- It is good practice to record when all network events occurred, for example valve closures. So that thorough analysis, if required, can occur at a later date.

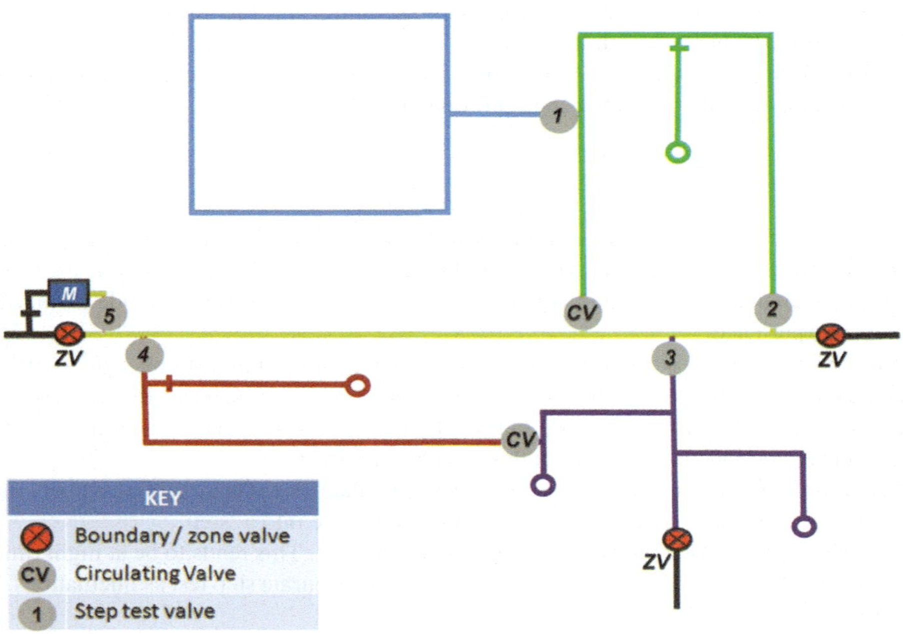

Figure 2.12 Showing a step test plan with valves labeled accordingly (*Source*: Primayer).

In Figure 2.13, drops in flow are shown as the 'steps' are being closed. The burst was located in the final step where the flow reduces considerably. The turning on of the 'steps' can be seen to the right hand side of the graph (Figure 2.13).

Figure 2.13 Graph showing results of a step test (*Source*: MAST PC software V5.03, Halma Water Management, Cwmbran, Wales).

The data should also be recorded manually so the data obtained from the graphs can be understood and this data can be analyzed after the event and confirmed that the reduction in flow was the result of a valve being turned off at the said time (Figure 2.14). It should also be kept as the audit trail of the test and the confirmation as to the findings.

Step Test Valving Schedule

DMA Name	Test Area		DMA Number	555
Test Details:				
Valve Number	Operation	Time	Flow(l/s)	Comments
Circ 1	Shut	02:00	10	
Circ 2	Shut	02:15	10	
Test 1	Shut	02:20	8	
Test 1	Open	02:25	10	
Test 2	Shut	02:37	9	
Test 3	Shut	02:51	5	

Figure 2.14 Table showing a traditional method of recording step test events (*Source*: Halma Water Management).

Data from the step test can now best sent to the operator using many ways including by GPRS means and the results from the test are displayed the same way seeing flows being reduced on each valve being turned off or step (Figure 2.15).

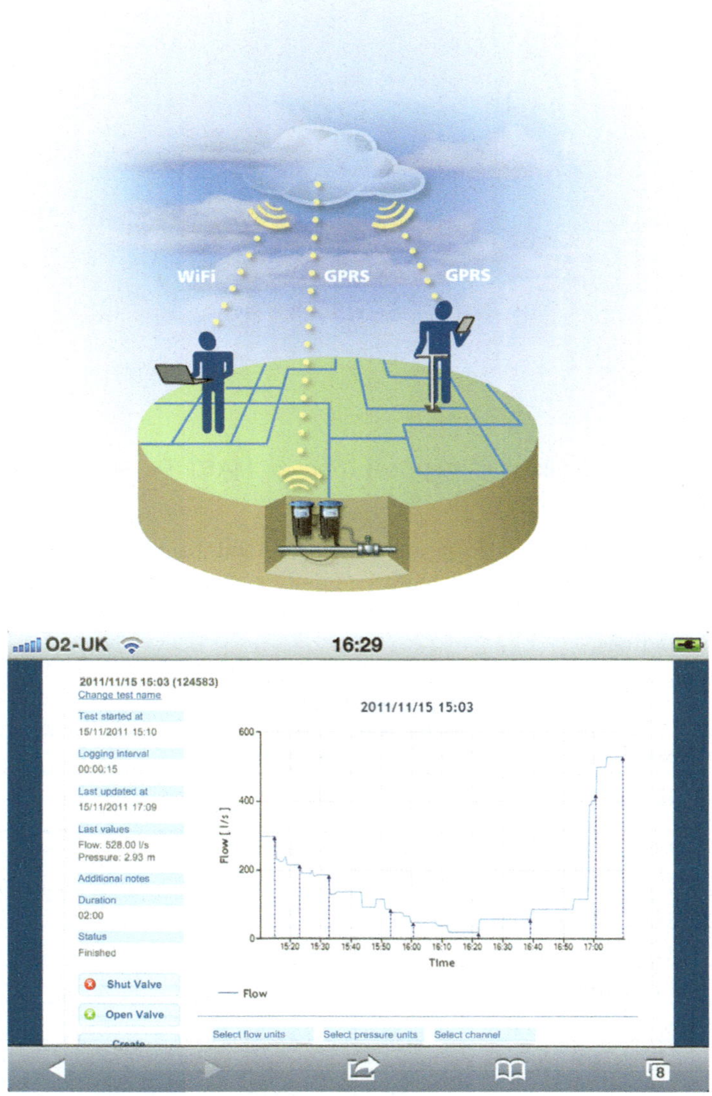

Figure 2.15 Advanced step testing (top) and data from advanced step test (bottom) (*Source*: Primayer).

Chapter 3

Principles of pressure management

3.1 INTRODUCTION

Pressure management is one of the most important WDM interventions that should be considered when attempting to drive down water losses. Leakage is driven by pressure and while it must be acknowledged that pressure management is not the answer in every case, it is often one of the most cost effective measures to reduce leakage and wastage that can be considered.

Pressure management can take many forms, ranging from the basic fixed outlet pressure control to some form of more sophisticated hydraulic or electronic control, which is often referred to as 'smart control' or 'advanced pressure control'.

Basic forms of pressure management have been in operation for many decades and the use of pressure reducing valves to provide fixed outlet pressure control is common practice throughout the world. Various enhancements have been developed over the years by different valve manufacturers to provide additional control and flexibility as it became apparent that a stable fixed outlet pressure into the zone was often not always the best solution. Electronic pressure controllers were first introduced in the early 1990s which provided greater control in allowing the outlet pressure from the PRV to be adjusted in accordance with either time of day or the flow requirement through the valve. Since then, various additional enhancements have been developed, either through more sophisticated electronic devices or through the numerous new hydraulic controllers. Each device has its benefits and limitations and the problem facing the water manager is often to select the most appropriate form of control to suit a specific application. There is unfortunately no 'one-size-fits-all' device when it comes to pressure management and the remainder of this section is devoted to discussing the most common forms of pressure control, together with their strengths and weaknesses.

3.1.1 Basics of pressure management

Water supply systems worldwide are generally designed to provide water to consumers at some agreed level of service, which is often defined as a minimum level of pressure at the critical point, which is the point of lowest pressure in the system. The typical average system pressure is usually considered to be in the order of 50 m with a minimum pressure of around 20 m. Many countries have recently introduced legislation restricting the maximum pressure and the limit is usually around 90 m, above which measures to reduce the pressure are required. A pressure of 90 m is actually very high and a more appropriate upper bound is 60 m which is now being adopted by many municipalities around the world in an attempt to reduce leakage and prolong the useful life of the reticulation system.

In the past 10 to 20 years, there has been a growing realisation that high water pressures are a more serious problem than previously thought. Not only are high water pressures responsible for excessive levels of leakage and new burst pipes, but they can significantly reduce the useful life of the reticulation system and the associated costs are rarely taken into account in the financial assessment of pressure management interventions. Unfortunately, there are generally very strict and specific regulations regarding water pressures in reticulation systems, which are essentially driven by the fire-fighting requirements. Deeper investigations often reveal that the fire fighting requirements are based on theoretical pressure requirements dating back to Victorian times. Few, if any, challenge these rules and regulations since to operate the systems below the fire-fighting requirements can result in protracted legal battles in the event that a fire does break out and the insurance companies try to apportion blame to anyone they can find in order to spread their liability. The threat of legal action in the event of a fire is often the underlying reason why many systems are operated at higher pressures than required.

Most water reticulation systems are designed to accommodate these pressure and flow requirements during the period of peak demand which would normally occur at some specific time of the day and during a particular month in the year. In other words, the systems are designed to provide the appropriate supply volumes and pressure during a very short period in the year and for the remainder of the time the systems tend to operate at pressures significantly higher than required. Even within the same system, there will be areas of high pressure due to topography and/or distance from the supply point with the result that many parts of a supply area will operate at pressures significantly higher than required in order to ensure that there is sufficient pressure at the one critical point.

Managing water pressures in a supply area is not a simple issue and there are a great many items to consider. The common factor in every system is the fact that leakage is driven by pressure and if the pressure is increased, the leakage will also increase.

In order to reduce leakage through pressure management it is necessary to reduce the water pressure without compromising the level of service with regard to the consumers and fire-fighting. As mentioned previously, most systems are designed to provide a certain minimum level of service in the system during the peak demand period, as shown in Figure 3.1. If the water pressure in a system can be reduced, even, for a short period during times of low demand, the water leakage from the system will be reduced. In this example, it is assumed that a minimum pressure of 20 m is required.

During the off-peak periods, which tend to be much greater than the peak periods, the system operates at a water pressure which is significantly higher than necessary, as shown in Figure 3.2. In effect, there are long periods when there is significant scope for pressure reduction and this is the basis on which the pressure management interventions are designed.

Reducing the water pressure in a system can be achieved in many ways ranging from simple fixed outlet pressure control to a variety of electronic or hydraulic controllers. Over the past 15 years, many new control devices have been developed, some of which offer greater sophistication and intelligence than previous devices. Smart controllers have become smarter with several of the latest devices incorporating artificial intelligence and various feedback loops to provide greater stability at the critical point.

Without detracting from the latest advances in pressure management, it is critically important for water supply managers not to lose focus on the real problem issues, as no piece of equipment, hydraulic or electronic, can replace the need for a properly designed, well managed and well maintained network. Ensuring that boundary valves are closed or open on a regular basis as per the network design is of greater importance than any pressure controller, which in any event will not operate properly when a pressure management zone has been compromised by unauthorized opening or closing of boundary valves.

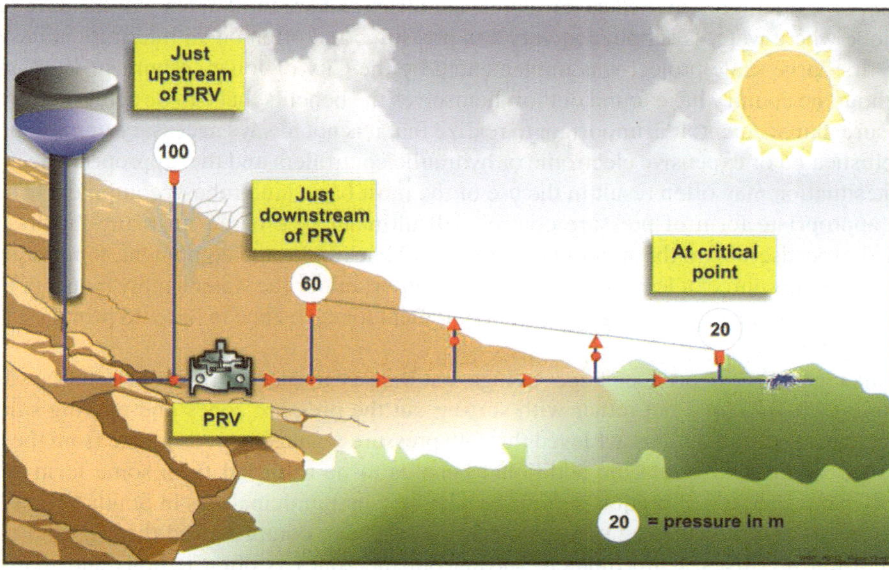

Figure 3.1 Typical pressure during peak demand periods.

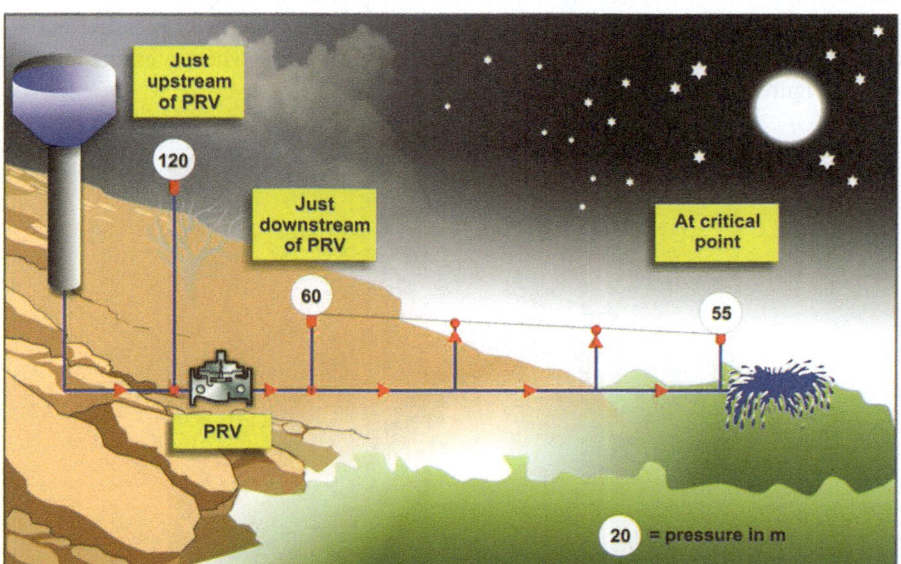

Figure 3.2 Typical pressure during off-peak periods.

South Africa is one of the most progressive countries in the world with regard to the use and implementation of many forms of advanced pressure control. It is one of the few countries in the world that includes both developed areas and developing areas side by side. Water pressures throughout the country tend to be similar if not higher than most developed countries and are unlike many of the developing areas across the globe where water is supplied at very low pressures and often on an intermittent basis.

Since the first large scale project was implemented by the City of Johannesburg in 1998, many other cities throughout the country have found out for themselves the benefits that can be achieved through some form of pressure management. It is important to realize that it is not always necessary or appropriate to use the most sophisticated or expensive electronic or hydraulic controllers and the 'appropriate technology' to suit a specific situation may often result in the use of the most basic (and robust) equipment. The selection of the most appropriate form of pressure control will ultimately be based on many factors, including, cost, technical expertise within the water utility, technical backup for the equipment, topography, leakage levels, as well as main source of leakage. It is therefore important for the water supply manager to carefully select the most appropriate pressure management equipment for each zone in order to provide a sustainable and effective solution.

Pressure management can produce huge savings, but the key to success rests not with the valves or the associated high-tech 'add-ons' but rather with sorting out the pressure zones and making sure that they remain discreet. Most of the savings achieved through pressure management originate from the basic fixed outlet pressure reducing valve and the additional benefits to be achieved from some form of advanced control are often exaggerated. The most serious problem facing municipalities in South Africa is ensuring that the pressure zones are operating properly and have not been compromised through an unauthorized opening or closing of valves. Unfortunately sorting out the zones is often tedious and thankless work which attracts little attention and is rarely the sort of material that makes a good conference paper. It is, however, the key to any successful pressure management installation.

It must always be remembered that pressure management does not eliminate leaks or repair any existing leaks. All leaks in the network before the introduction of pressure management remain as leaks after the pressure management installation has been commissioned. The benefits of pressure management are mainly through the fact that the leaks will run at a lower rate at any time where pressure has been lowered, as can be seen in Figure 3.3 and Figure 3.4.

Figure 3.3 Underground leak running at low pressure (*Source*: Ken Brothers).

Figure 3.4 Underground leak running at high pressure (*Source*: Ken Brothers).

In addition, the lower pressure regime will lower the rate at which new leaks develop and in many cases the incidence of new bursts can be reduced significantly – sometimes 90% or even more. The financial benefits of the reduction in new bursts and prolonging the life of the reticulation system are rarely included in the financial analyses despite the fact that they dominate the financial viability of most pressure management projects. Pressure management will usually provide a very attractive pay-back on the initial investment.

3.1.2 Concepts of pressure management

From experience, the most important issue when trying to introduce any form of pressure management is ensuring that the zone being considered is discrete and remains discrete. Only then, can the pressure management installation operate properly. Installing and operating controllers is the easy part of this intervention while establishing a proper pressure zone is often the most difficult and time consuming part of the work. If the zones are not discrete and are interlinked in any way, then pressure management is unlikely to function properly and should not even be considered until the underlying zone problems have been resolved. Water supply systems worldwide are generally designed to provide water to consumers at some agreed level of service, which is often defined as a minimum level of pressure at the critical point, which is the point of lowest pressure in the system. In addition, there may be certain fire-flow requirements which can over-ride the normal consumer requirements. The systems are designed to accommodate these pressure and flow requirements during the period of peak demand, which would normally occur at some specific time of the day and during a particular month in the year. In other words, the systems are designed to provide the appropriate supply during a very short period in the year and for the remainder of the time, the systems tend to operate at pressures significantly higher than required. Even within the same system, there will be areas of high pressure due to topography and/or distance from the supply point with the result that many parts of a supply area will operate at pressures significantly higher than required in order to ensure that there is sufficient pressure at the one critical point.

Managing water pressures in a supply area is not a simple issue and there are a great many items to consider. The common factor in every system is the fact that leakage is driven by pressure and if the pressure is increased, the leakage will also increase. If the water pressure in a system can be reduced, even, for a short period during times of low demand, the water leakage from the system will be reduced.

Many theories have been postulated to explain the pressure–leakage relationships in a municipal water supply system with the FAVAD (fixed area variable area) being the most widely accepted approach. Contrary to popular belief worldwide, the origin of the FAVAD theory can be traced back to a paper by Ledochowski (1956) who completed his research in Johannesburg. It is therefore very fitting that Johannesburg should have been the first city to implement advanced pressure control on a relatively large scale in South Africa back in 1998.

In short, the relationship between pressure and leakage will conform to a square-root relationship in cases where the size of the leakage path (i.e., hole) remains constant during the change in pressure. This is the typical situation when the leak is a small hole in an iron or steel pipe (i.e., a fixed area leak) in which case doubling pressure will result in approximately a 41% increase in leakage. In the case of leaks from plastic pipes or from cracks in asbestos cement pipes, the surface area of the leakage path does not remain constant when the pressure changes and such leaks will often open up to create a larger hole through which the water can leak. Such leaks are referred to as variable area leaks and if the pressure is doubled, the leakage will increase significantly more than from a fixed area leak. In some cases, the leakage will increase by as much as three times the original level.

In most systems, there tends to be a mixture of fixed area and variable area leaks and the split will depend on the proportion of steel/iron pipes to plastic/asbestos pipes. Considerable research and many papers have been presented on this topic in which different formulae are suggested to predict the impact of changes in pressure on leakage. From the author's experience, it is found that certain other factors often play a more critical role in the pressure–leakage relationship. For example, it has been found in many parts of South Africa that the quality of workmanship when laying the pipes is one of the most important factors influencing the eventual leakage from the system. Two similar systems next to each other can have significantly different leakage characteristics simply because one was laid properly with adequate site supervision while the other was laid by a poorly qualified contractor with poor supervision – usually to save on cost. In such cases, there is no adequate theory to explain the different responses to changes in pressure.

In order to reduce leakage through pressure management, it is necessary to reduce the water pressure without compromising the level of service with regard to the consumers and fire-fighting. Reducing the water pressure in a system can be achieved in a number of ways each of which has advantages and disadvantages. The following techniques are discussed:

- Fixed outlet pressure control
- Time-modulated pressure control
- Flow modulated pressure control
- Closed loop and hybrid control.

3.1.3 Fixed outlet control

Fixed outlet pressure control involves the use of a device, normally a pressure reducing valve (PRV) which is used to control the maximum pressure entering a zone, as can be seen in Figure 3.5. This is possibly the simplest and most straightforward form of pressure management as it involves the use of a PRV with no additional equipment. The advantages of this form of pressure control are:

- It is relatively simple to install as it requires only a PRV;
- Cost is lower as it involves no additional hydraulic or electronic controllers;
- Maintenance and operation is relatively simple.

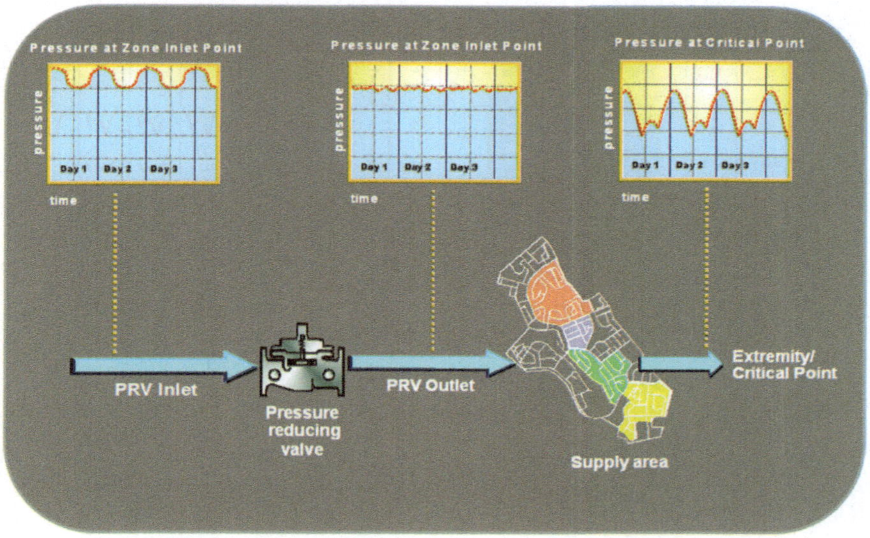

Figure 3.5 Fixed outlet pressure control.

The main disadvantage of simple fixed-outlet pressure control is the lack of flexibility to provide variable water pressures at different times of the day with the result that the maximum possible savings cannot be achieved. In many instances, fixed outlet pressure control is the preferred option due to its simplicity and the fact that the maintenance teams may not be able to operate and maintain the more sophisticated electronic equipment required when using some form of advanced pressure control.

3.1.4 Time control

The time-modulated pressure management option is shown in Figure 3.6 and is effectively the same as the fixed-outlet system with an additional device which can provide a further reduction in pressure during off-peak periods. This form of pressure control is useful in areas where water pressures build up during the off-peak periods – typically during the night when most of the consumers are asleep. The main advantages of this option are:

- The controller provides greater flexibility by allowing pressures to be reduced at specific times of the day, resulting in greater savings;
- The electronic controller is relatively inexpensive;
- The controller is relatively easy to set up and operate;
- The installation does not require a flow meter as the controller connects directly to the pilot on the PRV.

The main disadvantage of time-modulated control is that it does not react to the demand for water and this can be a problem if a fire breaks out requiring full pressure for fire-fighting. This problem can be overcome to some extent through the installation of a flow meter. In addition, the time-modulated option is more expensive than the fixed outlet option and does require a higher level of expertise to operate and maintain the installations.

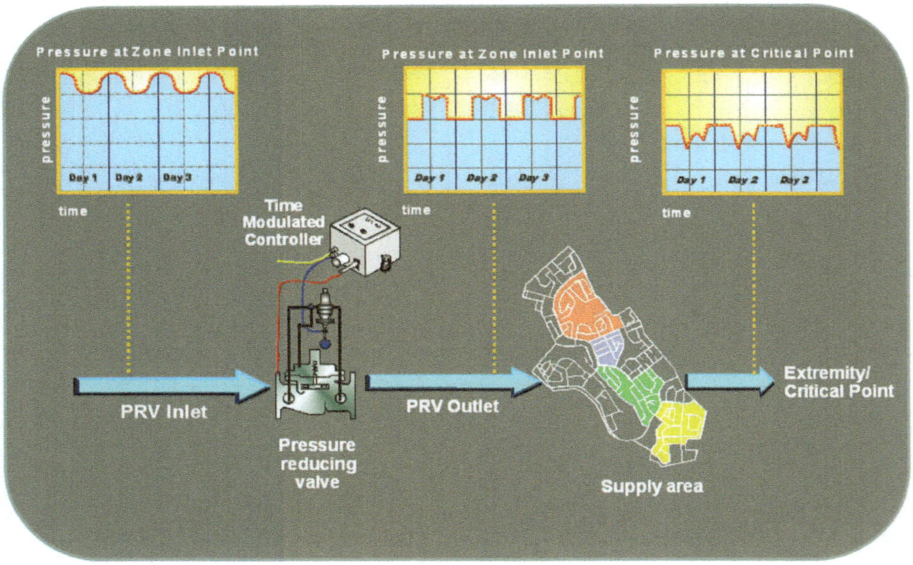

Figure 3.6 Time-modulated pressure control.

A typical time-control installation is shown in Figure 3.7, from which it can be seen that the controller is only connected to the PRV via three coloured plastic pipes (red, green and blue). The fourth cable which is connected to the meter is an optional cable that is simply used to log the flow going through the installation but has no influence on the controller. The controller itself, is a simple device which is easily set up using the two selection buttons on the front panel to set up the time for switching between the high and low pressure settings. A simple round knob-like dial on the bottom of the controller is used to control the low pressure setting while the high pressure setting is fixed by the pilot on the PRV. Should the controller fail for any reason, the PRV will revert to the outlet pressure setting as controlled by the pilot of the PRV. It should be noted that there are a number of similar controllers available in the market and they all have slightly different features.

3.1.5 Flow control

Flow modulated pressure control as shown in Figure 3.8 provides even greater control and flexibility than time control. It will normally provide greater savings than either of the two previously mentioned options but this greater flexibility (and savings) comes at a price. The electronic controller is more expensive and it requires a properly sized meter in addition to the PRV. It may not always be cost effective to use the flow modulated option and careful consideration should be given to the specific application before selecting flow-modulated control. One key advantage is that the flow modulated option will not hamper the water supply in the case of a fire but the additional savings achieved may be offset against the extra cost of the controller and need for a meter, together with the likely additional down-time associated with a more complicated device. There are more components in a flow modulated installation which, at some time, are likely to fail. The additional savings generated from such a controller must therefore be weighed up against the likelihood of such component failures and the availability of the necessary technical support to operate and maintain the equipment.

Figure 3.7 Typical PRV with time-control (*Photo*: Niel Meyer).

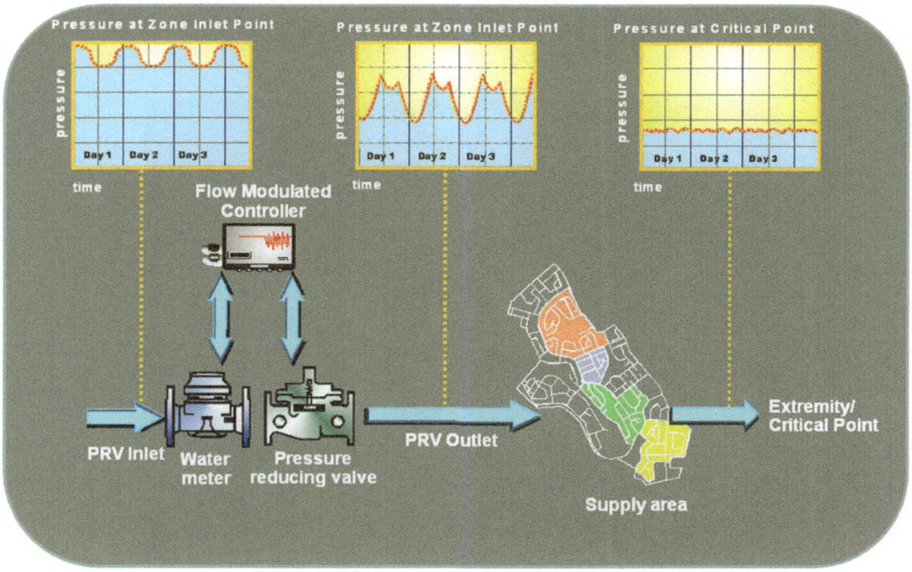

Figure 3.8 Flow modulated pressure control.

A typical example of flow modulated pressure control is shown in Figure 3.9 from which it can be seen that the downstream pressure profile closely mimics the flow through the installation.

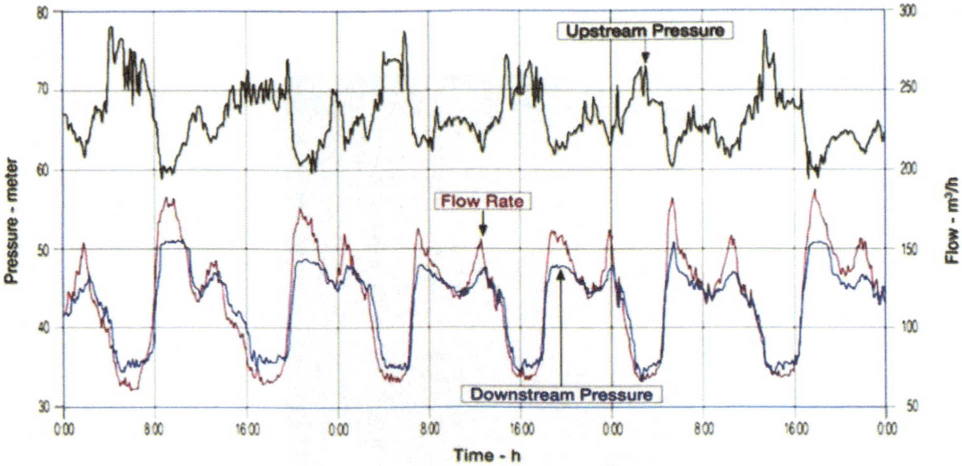

Figure 3.9 Flow modulated pressure control.

This form of pressure control has various advantages over the simpler time-control or fixed outlet control in that it provides greater flexibility and will generally achieve greater savings than the simpler forms of control – usually an additional 10% to 20%. It also helps to address any concerns regarding the fire-fighting flow requirements in the respect that in the case of a fire, the controller will open up if required to maintain the necessary system pressure to support the increased flow.

Figure 3.10 shows a typical set-up for a flow modulated controller in which the key items necessary to achieve flow control are indicated.

Figure 3.10 Typical flow modulated controller set up.

3.1.6 Intelligent and closed-loop control

In most cases, flow modulated pressure control is as sophisticated as most water utilities will consider. In many instances, the use of flow modulated control is beyond the human resource skills base of the water utility and as a result time-modulated control or basic fixed outlet pressure control is the preferred and most appropriate option. In recent years, however, several more advanced 'closed-loop' and 'self-learning' systems have been implemented in various parts of the world which claim even greater savings than those achieved from flow control with claims of an extra 2% to 10% over standard flow control. In these more advanced installations, a pressure sensor at the critical point is generally used to monitor the pressure and provide some form of feedback to the controller at pre-defined intervals. The feedback to the controller can be on a real-time basis in which case it will normally update the controller on the pressure status at the critical point every 15 minutes or so. In other cases, it will record a full day of pressure information before transmitting the data to the controller, which will then adjust the pressure profile for the following day. These forms of pressure control as shown in Figure 3.11 and Figure 3.12 can provide the near ultimate level of control and therefore also the maximum savings that can be achieved.

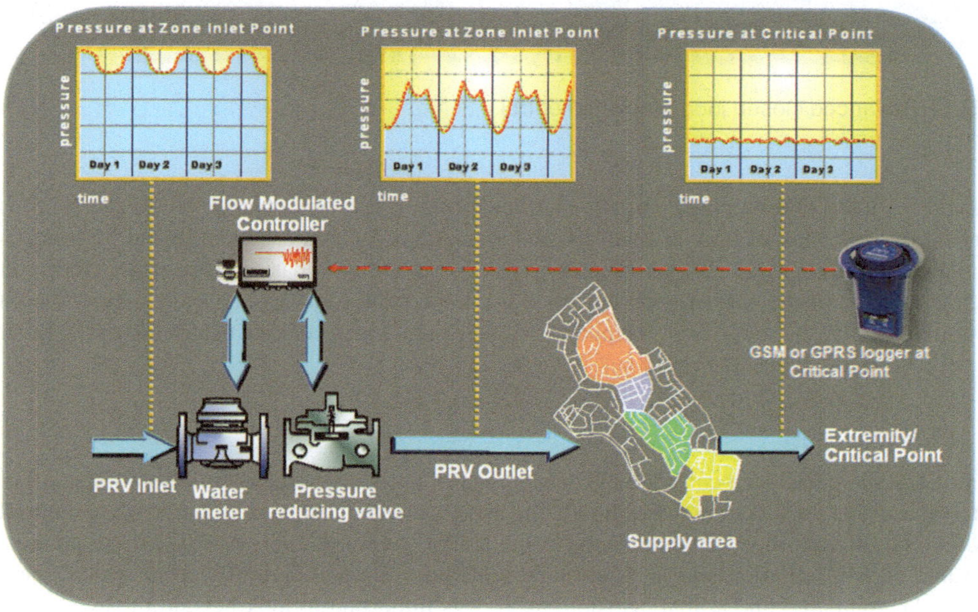

Figure 3.11 Closed loop pressure control.

Once again, both approaches have certain advantages and disadvantages which should always be taken into account when trying to establish the most appropriate form of pressure control for a specific application. Any additional savings will come at a cost both in monetary terms and also the technical expertise needed to support the long-term sustainability of the equipment and the possibility of component failure, which will happen at some time, as well as the battery life for the units in cases where they are battery powered.

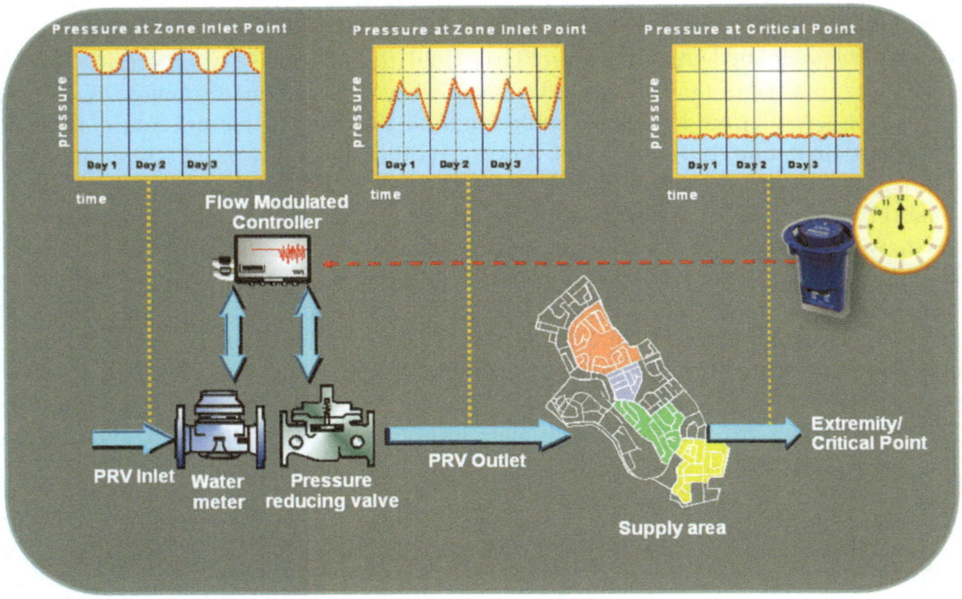

Figure 3.12 Advanced closed loop pressure control.

It should be noted that there are numerous other forms of pressure control which can be considered when trying to reduce losses from a water distribution system. Several excellent hydraulic controllers are also available in addition to a number of simpler devices based on a basic orifice plate. No-matter what approach is selected, if the water pressure in an area can be reduced for even a short period, there will be a reduction in leakage, a reduction in the number of new burst pipes, and the useful life of the reticulation system will be extended to some degree.

3.1.7 When to use advanced pressure control

In recent years, there has been considerable debate regarding the use of electronic controllers on pressure reducing valves. The most common question being asked is 'when is it appropriate to use a controller and if so, what type of controller'. Unfortunately, in the author's opinion, there is no clear cut answer to this question and each installation and pressure zone must be examined carefully, after which, the various options can be compared and a final decision taken. It is important to look beyond the hype and spin of some suppliers who promote the use of controllers in virtually every case when in reality, many situations do not warrant the use of anything more than a standard pressure reducing valve. Based on the results achieved from over 400 pressure management installations, it appears that the largest saving achieved from a pressure management installation is usually from the basic fixed outlet pressure control which simply involves a pressure reducing valve and associated pilot. In most cases, this will provide around 60% to 70% of the ultimate savings that can be achieved and in some cases even more. Obviously, the benefits will vary from case to case and will depend on the characteristics of the pressure zone in question. Adding a time-control can often give a further 10%–20% – again depending on the characteristics of the zone. A flow controller will bring up the ultimate savings to between 90% and 95% of the ultimate savings and some form of closed-loop or self-learning control will provide the remainder. Any claims that a controller is directly responsible for generating 60% of the savings or more, are false and tend to include the combined benefits of the fixed outlet control and the controller.

In summary, if a zone experiences large pressure fluctuations at the critical point between peak demand and off-peak demand periods, there will often be scope for additional savings through the use of a controller. In other cases, where a system does not experience major pressure fluctuations at the critical point between the peak and off-peak periods, there will be little benefit achieved through the use of a controller unless the client is willing to accept a different critical point pressure at night compared to that required during the day.

To highlight the above issue, the Eerste River Case Study from the City of Cape Town provides an excellent case study in which a controller is not considered necessary. The Eerste River water supply system initially experienced very high water pressures in excess of 80 m, resulting in high water losses through infrastructure leaks and a high incidence of burst pipes. The level of leakage at the start of the project before any form of pressure management was introduced can be seen in Figure 3.13, which indicates an average daily flow of ±265 m^3/hr and a minimum night flow of ±158 m^3/hr. The ratio of MNF to ADD (often used as a rough indicator of leakage levels) of ±60% was very high and suggested very high levels of leakage in the area – a target value of around 20% would normally be considered acceptable. As a result of the high water pressure, the zone experienced approximately 240 bursts per annum before any form of pressure control was introduced. This naturally had major cost implications as well as service delivery implications.

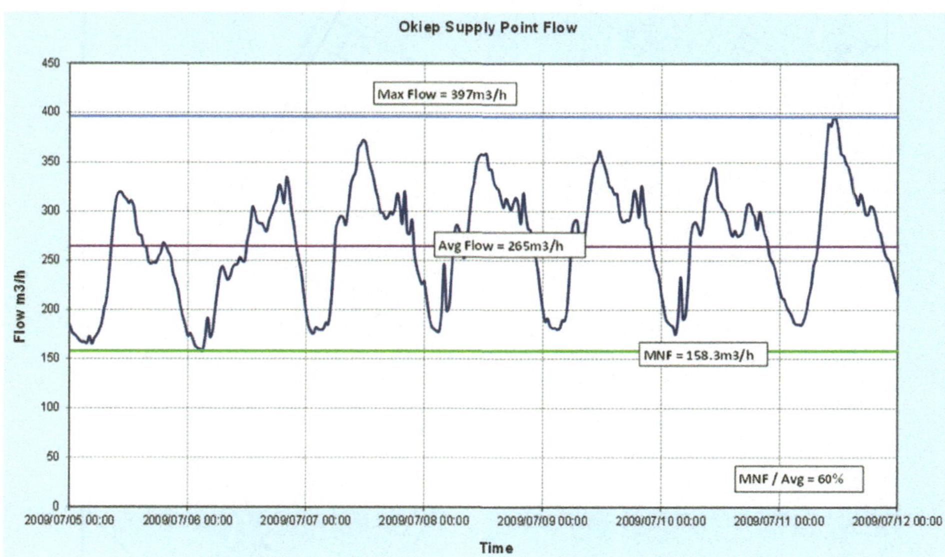

Figure 3.13 Original flows into Eerste River.

To address this problem, the City of Cape Town commissioned the design and implementation of a single pressure management installation on the main supply pipeline into the area. Before introducing the pressure management, it was first necessary to sectorize the area and eliminate all bottlenecks caused by open or closed valves, which first had to be identified. This is often the most difficult and time-consuming part of any pressure management initiative and only after the zone has been properly sectorized can the pressure be reduced.

The area involved is shown in Figure 3.14 and the PRV installation involved a single chamber housing a 250 mm diameter valve. The initial pressures in the area are shown in Figure 3.15, which highlights that the pressures at both the inlet and critical point were very high, with critical point pressures of between 60 m and 80 m. In such an area, there is clearly significant scope for some form of pressure management and the main PRV installation was completed in 2009. It is also important to note that in this specific case, the variation in pressure at the critical point between peak and off-peak demand periods was relatively small. This is unusual as it indicates that the reticulation network is under relatively little stress during the peak demand periods and often indicates that the area has either yet to achieve its full developed demand or that the main supply pipes are oversized to some degree.

Figure 3.14 Map of the Eerste River pressure management zone.

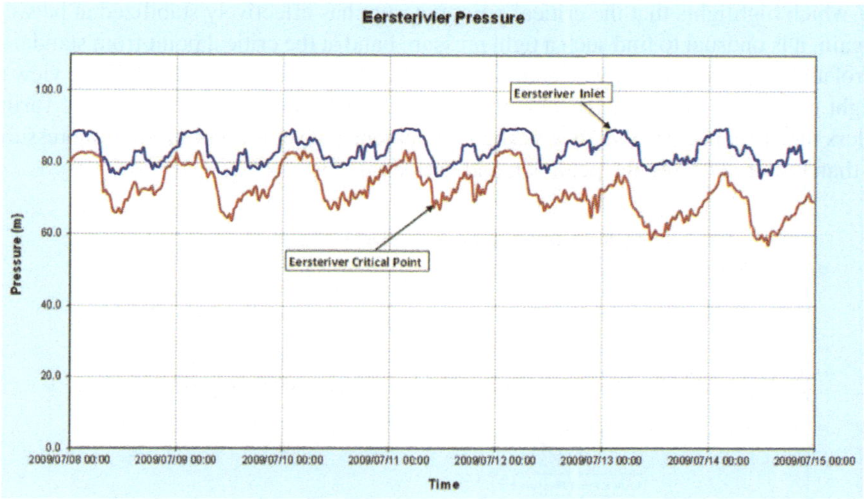

Figure 3.15 Original pressure in Eerste River.

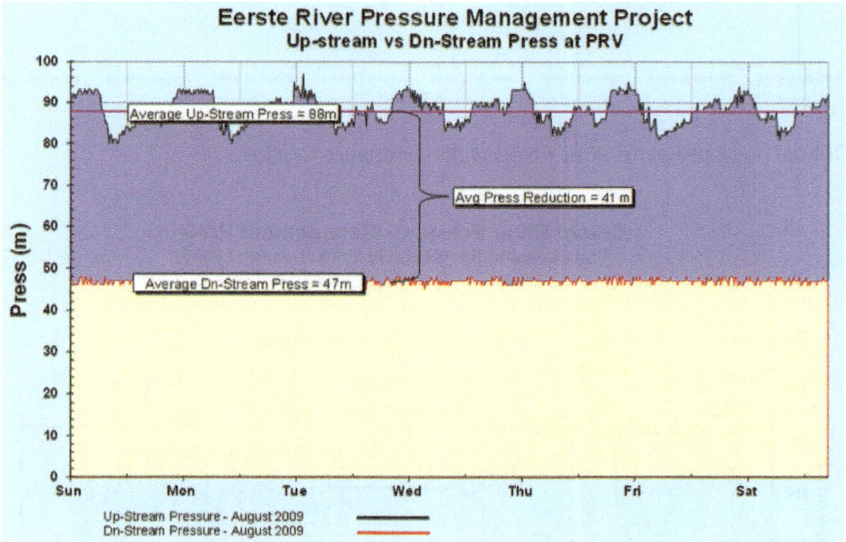

Figure 3.16 Pressures after implementation of Fixed Outlet Pressure Control.

Before even considering the use of a controller, it is recommended that the zone is properly sectorized and monitored using the basic PRV based on fixed outlet pressure control. This is the first and most important aspect of the commissioning process since a controller or some other form of advanced pressure control should only be considered once the basics are firmly in place and the zone is operating properly. In this case, the pressure at the outlet of the PRV was reduced from over 80 m to approximately 47 m, as shown in Figure 3.16, based on the client's request that the pressure at the critical point should not drop below 35 m. This can be seen in Figure 3.17 which highlights that the critical point pressure has effectively stabilized at between 38 m and 35 m. Once again, it is unusual to find such a tight pressure band at the critical point from standard fixed outlet pressure control and often this can only be achieved through the use of a controller. It is the view of the author that if such tight pressure control is achievable through a normal fixed outlet PRV, then no further control is necessary unless the client decides that they wish to implement a lower level of minimum pressure during off-peak periods than required during the peak demand periods.

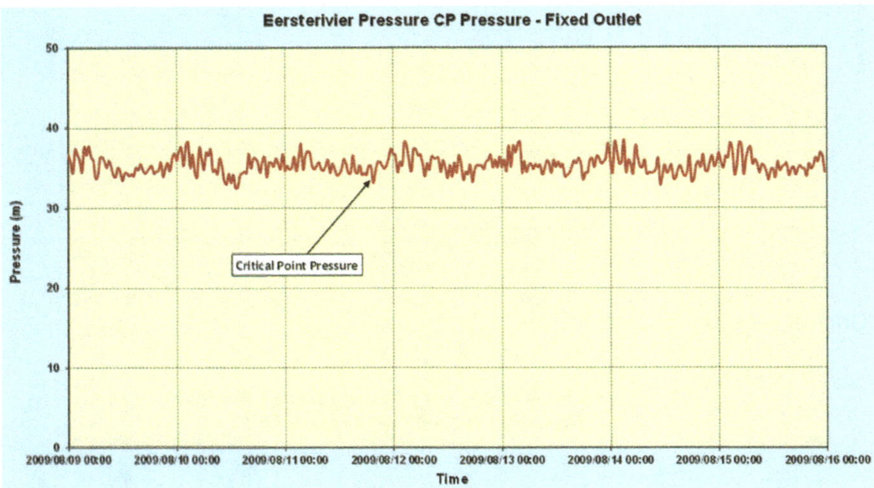

Figure 3.17 Critical point pressure after Fixed Outlet Pressure Control.

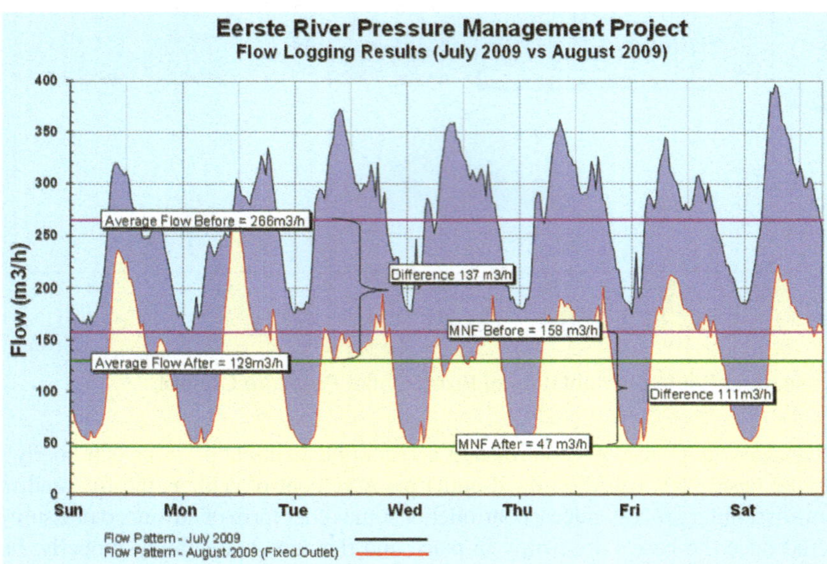

Figure 3.18 Flow savings from Fixed Outlet Pressure Control.

The net result of the reduction in pressure is shown in Figure 3.18 where the reduced water consumption is highlighted by the shaded area shown in blue. As can be seen in Figure 3.18, the savings are highly significant and clearly demonstrate the benefits that can be achieved from standard fixed outlet pressure control. The MNF dropped from ±158 m^3/hr to ±47 m^3/hr and there was still considerable scope for further pressure reduction since a minimum critical point pressure of 20 m is usually sufficient.

A summary of the initial savings generated through fixed outlet control at the Eerste River pressure management installation is provided in Table 3.1. The value of the water saved is based on a cost of water production of R6.20/m^3, as supplied by the City of Cape Town at the time of commissioning.

Table 3.1 Eerste River direct savings from fixed outlet pressure control.

Area	Savings (m^3/day)	Savings (m^3/year)	Value savings/ year @ R6.20/m^3	Implementation cost	Payback period
Eerste River	3288	1,200,120	R7,440,744	R2,000,000	0.3 years

As can be seen from the figures provided in Table 3.1, the payback for this specific installation was around 4 months based on the initial savings. The use of fixed outlet pressure control in this area resulted in a leakage reduction of ±70% based on the minimum night flow dropping from ±158 m^3/hr to ±47 m^3/hr. The total water use in the zone reduced from ±266 m^3/hr to ±129 m^3/hr representing a reduction of approximately 50%, as can be seen in Figure 3.18. In addition, the average number of burst pipes per annum was reduced from over 240 per annum to less than 24 per annum following the initial intervention – a drop of approximately 90%.

Returning to the initial question of whether or not to use a controller, the answer lies in Figure 3.17 which shows the new pressure at the critical point. This figure highlights that there remains scope for further pressure reduction in the area if it is acceptable to drop the minimum pressure to 20 m at the critical point. What is very interesting, however, is the fact that the pressure variation at the critical point is only around 5 m and there is relatively little additional benefit to be gained from any form of advanced pressure control. Some experts will insist that there is still a case for using a controller even in such areas if it can be shown that the additional savings can be justified from a cost/benefit analysis. Much will depend on the particular wishes of the client as from experience it is found that some clients prefer not to use any form of advanced pressure control while others seem to favour controllers in every case. In the view of the author, if the additional savings are not highly significant and a stable critical point pressure has been achieved using standard fixed outlet control, then a controller is not necessary.

In order to establish the additional benefits of further pressure control in this area, the client agreed to test the additional savings that could be achieved through further pressure reduction. It was agreed to reduce the fixed outlet pressure from 37 m to 27 m, which was achieved by simply lowering the fixed outlet pressure in the normal manner. An additional drop of 4 m for approximately 6 to 8 hours each night using an electronic controller. Based on the published results, the overall water use dropped to an average of 127 m^3/hr, the minimum night flow dropped to 32 m^3/hr and the number of new burst pipes dropped from around 24 per annum to around 12 per annum. Clearly the introduction of more advanced pressure control does result in additional savings and the only debate is to decide if the additional savings are sufficient to motivate the additional expense and associated maintenance associated with the controller. As always, there are two views on this and ultimately it is the decision of the client.

The previous example from Eerste River demonstrated a zone where significant savings could be achieved through the implementation of pressure management, although there was little additional benefit from the use of advanced pressure control due to the fact that the pressure at the critical point was already stable through the use of a normal fixed outlet PRV. In this second example, another zone is used to highlight a case where there is clearly greater scope for advanced pressure control. The example is again from the City of Cape Town in an area called Langa as shown in Figure 3.19.

Figure 3.19 Location plan for Langa.

The zone is not particularly large but makes an interesting case study due to the fact that the pressure at the critical point is highly variable unlike the previous example where there was very little difference between peak demand pressure and off-peak pressure. The inlet pressure, PRV outlet pressure and flow profile for Langa are shown in Figure 3.20, in which it can be seen that the PRV has been set to provide a fixed outlet pressure of ±52 m and the valve is indeed providing a stable and continuous pressure into the zone. The resulting pressure at the critical point is shown in Figure 3.21, which clearly highlights a significant pressure variation between peak conditions and off-peak conditions. Such a pressure variation is often observed in pressure zones and is a function of the characteristics of the reticulation system and the system demand. If the demand is high relative to the capacity of the water mains, then at peak demand conditions the friction losses will be high and as a result, the system pressure at the critical point will be significantly lower than that provided into the zone at the PRV. During off-peak conditions, there is little demand and therefore the friction losses are also low with the result that the pressure at the critical point is at or near the zone inlet pressure. This is clearly reflected in Figure 3.21.

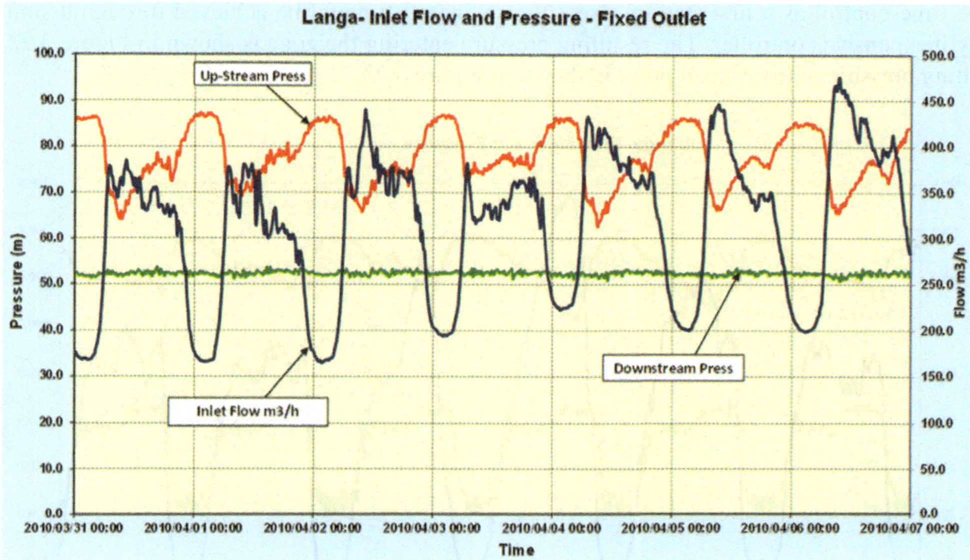

Figure 3.20 Inlet and PRV outlet pressure into Langa.

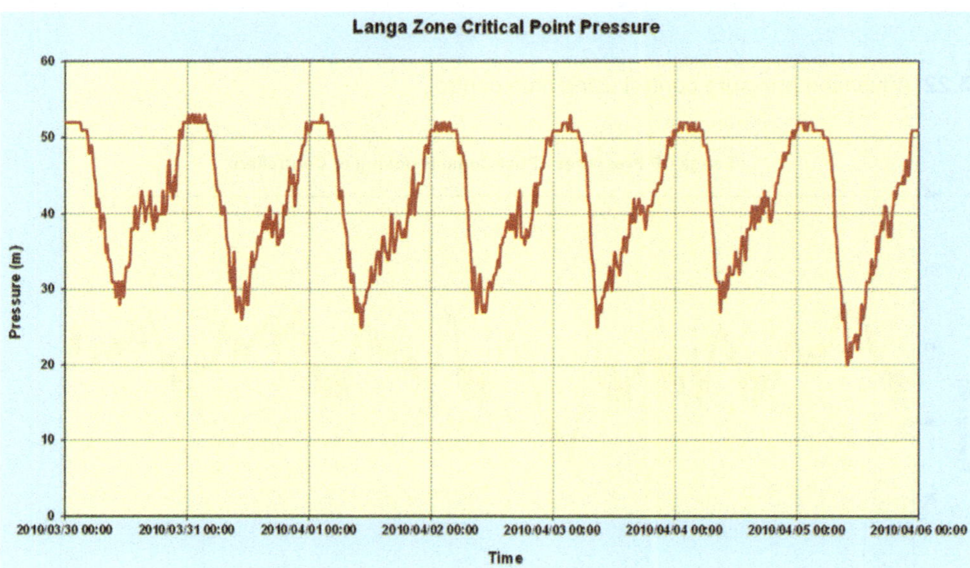

Figure 3.21 Pressure at the critical point in Langa.

Zones with such a high pressure variation at the critical point are often ideal candidate zones for some form of advanced pressure control. As always, the question is to select the most appropriate form of pressure control to suit the zone and the needs of the client. In this particular case, it was decided to use basic time-control as a first step to assess the savings that could be achieved through a simple and relatively inexpensive controller. The resulting pressure entering the zone is shown in Figure 3.22, while the resulting pressure at the critical point is shown in Figure 3.23.

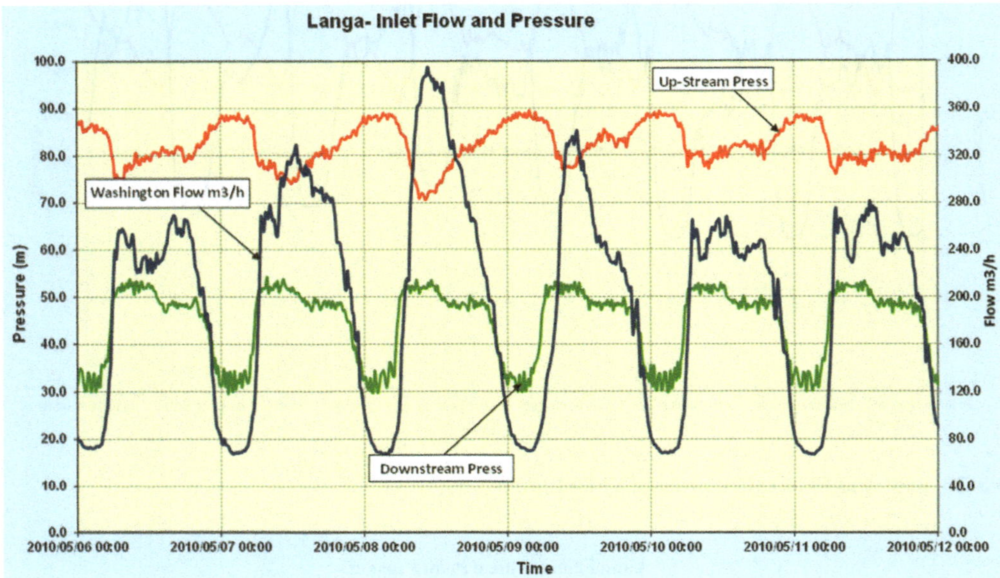

Figure 3.22 Advanced pressure control using time control.

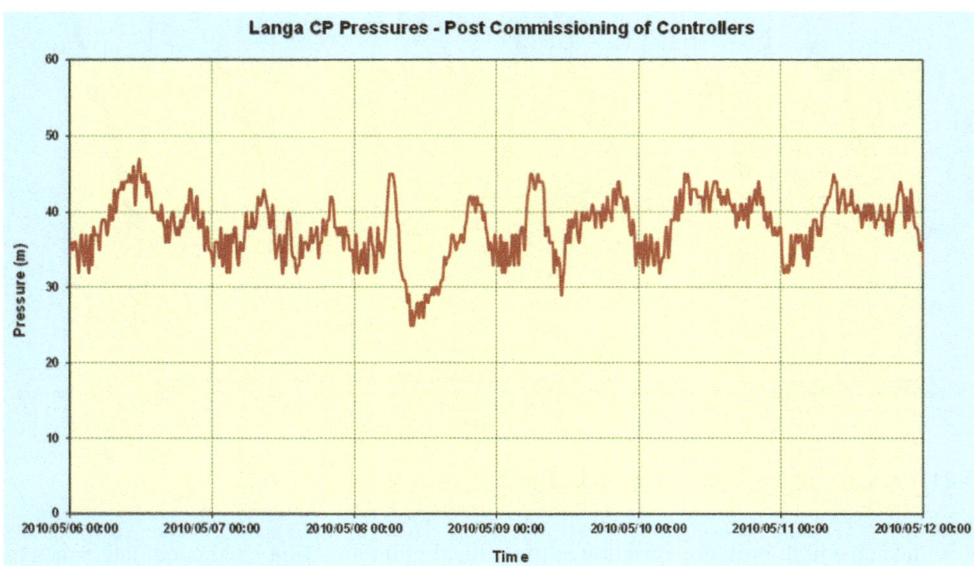

Figure 3.23 Critical point pressure using time control.

From Figure 3.23 it can be seen that the pressure variation at the critical point has been reduced significantly from the original situation, as shown previously in Figure 3.21. The leakage in the area has also been significantly reduced from ±200 m^3/hr when using the fixed outlet PRV to ±80 m^3/hr when lowering the night-time pressure from ±52 m^3/hr to ±30 m^3/hr. As can be seen in this example, the savings based on the minimum night flow analysis were very significant and in addition, there was a marked reduction in new burst pipes.

The pressure variation at the critical point is not yet stable and indicates a variation of ±10 m to ±15 m, which can be improved through the use of a more sophisticated (and expensive) pressure controller. The use of some form of flow control can be viable in this example and would most likely provide further savings of between ±10% and ±20%, however, after discussions with the client it was agreed to use the basic time control for a period and to review the situation after the residents had become used to the lower pressures in the area. The Langa zone clearly demonstrates that the additional control provided by an electronic controller is worthwhile in areas that experience high pressure fluctuations at the critical point between peak and off-peak demand periods.

3.1.8 Some large advanced pressure management installations

There are several hundred advanced pressure control installations in South Africa including at least three of the largest installations of their type in the world. While many of the smaller installations are of significance and of interest for a variety of reasons, the three largest installations are mentioned in this section since they have helped to create awareness both locally and internationally of the benefits that can be achieved through advanced pressure control. The three large installations to be discussed are Khayelitsha, Sebokeng and Mitchells Plain – all three of which remain fully operational at the time of writing this report and were 12, 8 and 4 years old respectively. Each of these installations is noteworthy in their own right for various reasons as discussed below.

3.1.9 Khayelitsha: City of Cape Town – 2001

Khayelitsha is one of the largest townships in South Africa and is located approximately 20 km from Cape Town on the Cape Flats, which is a large flat sandy area at or near sea level. There are ±43 000 serviced sites with both internal water supply and water borne sewage while there are a further ±27 000 low-cost housing units which are supplied from communal standpipes supporting a population of ±450 000.

At the beginning of 2000, the water supplied to Khayelitsha was measured to be ±22 million m^3/a. The level of leakage was estimated from the night-time water use to be almost three-quarters of the water supplied to the area. The Minimum Night flow (MNF) was measured to be in excess of 1600 m^3/hr which is sufficient to fill an Olympic-sized swimming pool every hour.

The main source of the leakage was identified as the household plumbing fittings which have been badly damaged through constant exposure to a relatively high pressures which could exceed 80 m at some times. Such leakage resulted in very high water consumption in most properties and high levels of non-payment since the customers could not afford to pay for new taps and toilet fittings, let alone the high water bills.

The Khayelitsha Pressure Management Project was commissioned in 2000 to improve the level of service to the Khayelitsha community by reducing the excessive water pressure and pressure fluctuations in the reticulation system. The layout of the installation is provided in Figure 3.24.

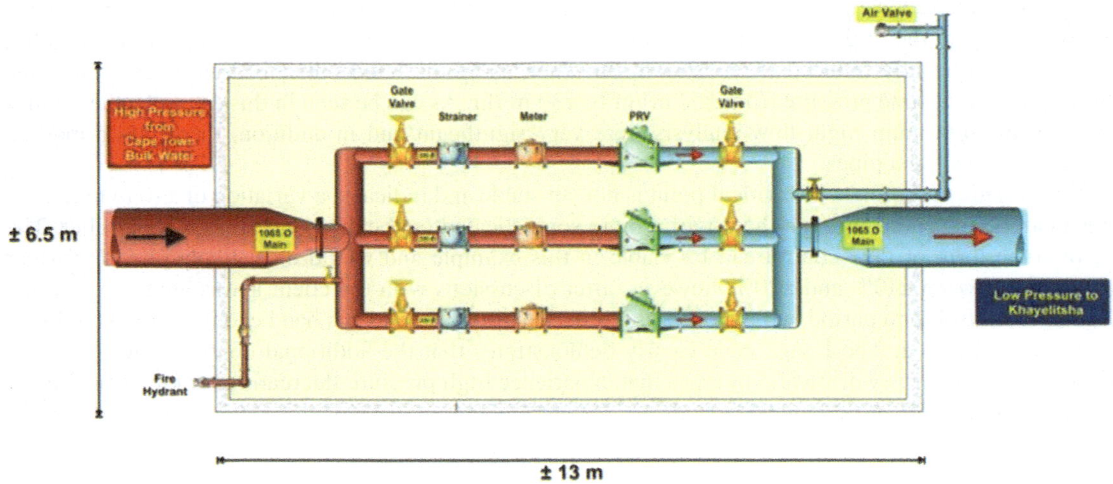

Figure 3.24 Schematic layout of Khayelitsha installation.

The average daily flow was reduced from ±2500 m³/hr to ±1500 m³/hr representing an annual saving of ±9 million m³/hr or approximately 40% of the original water use. The minimum night flow was reduced from ±1600 m³/hr to ±750 m³/hr. The number of burst pipes reduced significantly, although these savings were not taken into account in the overall cost benefit analysis. Local labour was used throughout the project and the community support was a key factor in the successful implementation of the project, which was implemented without any incidences of theft or vandalism. The community awareness process was designed and implemented by the City of Cape Town using its own personnel in parallel with the project implementation.

A summary of the actual savings achieved from the first two years of operation is provided in Table 3.2 (Mckenzie *et al.*, 2003). It should be noted that the latest estimates of savings achieved from the installation made by the City of Cape Town suggest savings of ±9 million m³/a with a financial saving of ±R54 million per year (± $6 million/a) (Meyer *et al.*, 2009).

Table 3.2 Summary of Khayelitsha savings for initial 2-year period.

Description	Basis of calculation	Volume saved	Value of saving (R million)
Direct water savings in 2002	Based on R3.09/m³	9 million m³	27.8
Direct water saving in 2003	Based on R3.49/m³	9 million m³	31.4
Delay to infrastructure – 2 years	7% of R35 million/yr		4.9
Maintenance and replacement	R250, 000 per year		–0.5
Total saving over 2-year period			63.6

The completed installation is shown in Figure 3.25 and Figure 3.26.

Figure 3.25 Khayelitsha pressure management installation.

Figure 3.26 External view of the Khayelitsha pressure management installation.

3.1.10 Sebokeng: Emfuleni Local Municipality – 2005

Emfuleni Local Municipality is located to the south of Johannesburg in the industrial heartland of South Africa. The municipality supplies water to approximately 1.2 million residents of which ±450,000 are located in the Sebokeng and Evaton areas. The areas are predominantly low-income residential areas with ±70,000 household connections, each of which is supplied with an individual water supply as well as water borne sewage. The combination of low income coupled with high unemployment has resulted in a general deterioration of the internal plumbing fittings over a period of many years, causing high levels of leakage which is characterized by a minimum night flow of ±2800 m^3/hr. This is one of the highest minimum night flows recorded anywhere in the world and represents almost two Olympic sized swimming pools of water every hour during a period when demand for water should be minimal. It was estimated that the wastage in the area before the project was commissioned was ±80% of the water supplied to the area, which in turn represented an annual water bill of ±R120 million per year (±$12 million).

In 2004, the municipality commissioned one of the largest advanced pressure management installations in the world as the first phase of a long-term strategy to reduce wastage in the area. The project involved no financial input from the municipality and even the initial capital costs were provided in total by the Project Team. The project was, effectively, a small-scale Public–Private Partnership involving a simple risk–reward model, details of which are provided in Figure 3.27 (Mckenzie & Wegelin, 2005).

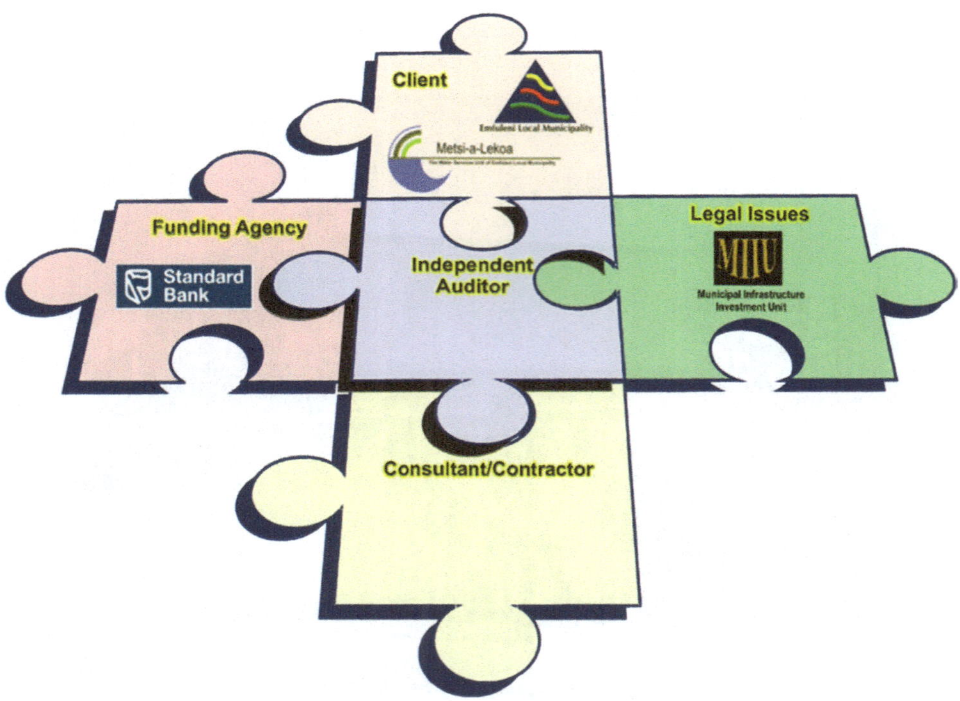

Figure 3.27 Organization framework for Sebokeng PPP.

The installation itself is shown in Figure 3.28 and is interesting for a number of reasons.

- The installation includes a manifold section at each end with five branch pipes connecting the two manifolds, which provides great operational flexibility.
- The middle pipe section has no PRV or meter and is used to recharge the downstream network, which may have emptied after a shutdown. If the supply pipes with meters are used to re-charge the network, there is a risk of damaging the meters in cases where mechanical meters are used. The original mechanical meters have now been replaced with magnetic flow meters to eliminate the meter problem,
- Valves are used in both manifolds to provide flexibility, both in allowing the pressure to be equalized between the two supply pipes which enter the chamber at different pressures due to their different pipe diameters and allowing the downstream pressure to be equalized if required. They also provide the opportunity to include the central pipe either to supply Sebokeng or Evaton, should this be necessary as they grow in future.
- All valves on the five supply pipes are resilient seal valves and not butterfly valves since they tend to provide a better seal in cases where valves are being opened and closed on a regular basis.
- Air valves are located on both manifolds and an enlarged collector tube is used to ensure that the air is in fact able to escape from the pipework. This is easier to see in Figure 3.29.

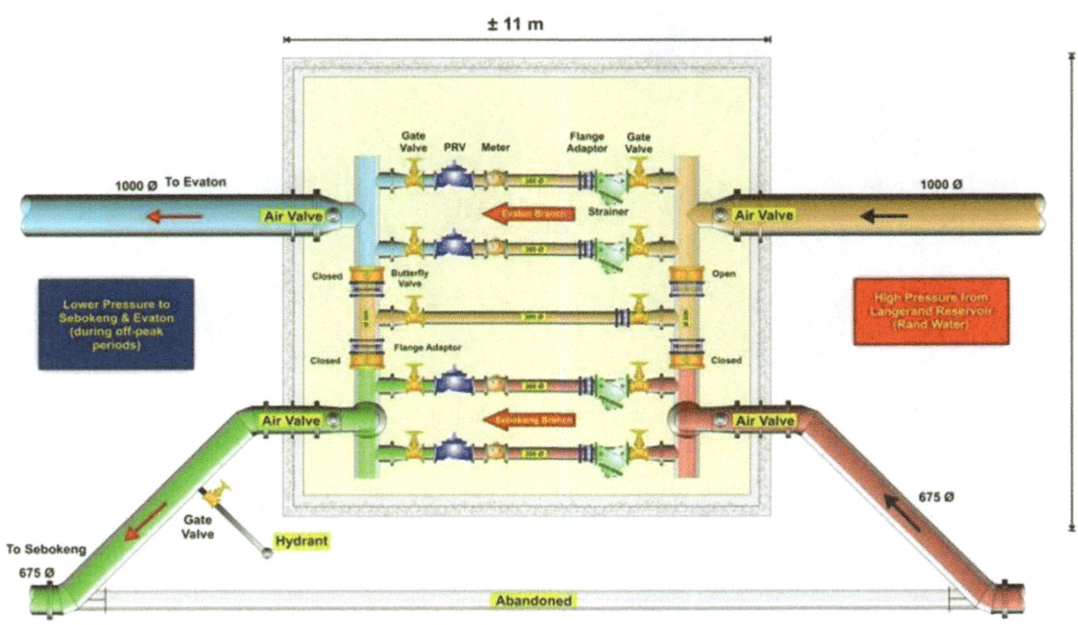

Figure 3.28 Schematic layout of Sebokeng installation.

The savings achieved in the first 30 months of operation of the installation exceeded all expectations of both the Project Team as well as the municipality and are the most obvious benefits to accrue from the project. After operating and managing the installation for several years, several other benefits also became apparent which were not initially anticipated. In particular, the following benefits have been achieved:

- Deferred upgrading of infrastructure
- Identification of bottlenecks in the system and problem infrastructure;
- Identification of bulk meter errors;
- Catalyst for funding;
- Improved municipality status;
- Creation of national WDM fund;
- Catalyst for other WDM interventions;
- Sustainability of savings.

Figure 3.29 Sebokeng/Evaton pressure management installation.

The project represents a significant advancement in Public–Private Partnerships (PPPs) and clearly demonstrates that small scale Public–Private Partnerships can be viable despite the general view that this type of project is confined to large-scale initiatives due to the effort and expense in developing the PPP type of contract. While the Sebokeng and Evaton Public–Private Partnership is clearly one of the most successful small scale PPPs to be completed in South Africa, the real benefits of the project are only now materializing many years after the project was commissioned. Both the project team and the municipality are very happy with the outcome of the project and are continuing to work together to build on the initial success. While the financial savings generated exceed all initial expectations, the hidden and often less tangible benefits greatly outweigh the obvious and tangible benefits.

The actual savings achieved are summarized in Table 3.3 and also depicted graphically in Figure 3.30. As can be seen from Figure 3.30, the water supplied to the area at the start of 2008 is almost the same as it was at the start of 2001, which clearly highlights the true level of savings that have been achieved.

Table 3.3 Summary of Sebokeng savings for first 42 months of operation.

Period	Water consumption (Kl)			Savings	
	Expected	Actual	Saving	Rands	US$
Months 1 to 6	18,721,000	14,614,000	4,107,000	11,499,600	1,437,450
Months 7 to 12	18,751,000	12,785,930	5,965,070	16,702,196	2,024,509
Months 13 to 18	19,403,000	13,886,451	5,516,549	16,218,654	1,908,077
Months 19 to 24	19,423,000	13,877,370	5,545,630	16,304,152	1,863,332
Months 25 to 30	20,086,000	15,269,040	4,816,960	14,788,067	1,643,119
Months 31 to 36	20,206,000	15,633,153	4,572,847	14,038,640	1,517,691
Months 37 to 42	20,827,000	15,870,850	4,956,150	16,107,488	1,695,525
Total	137,417,000	101,936,794	35,480,206	105,658,797	12,089,702

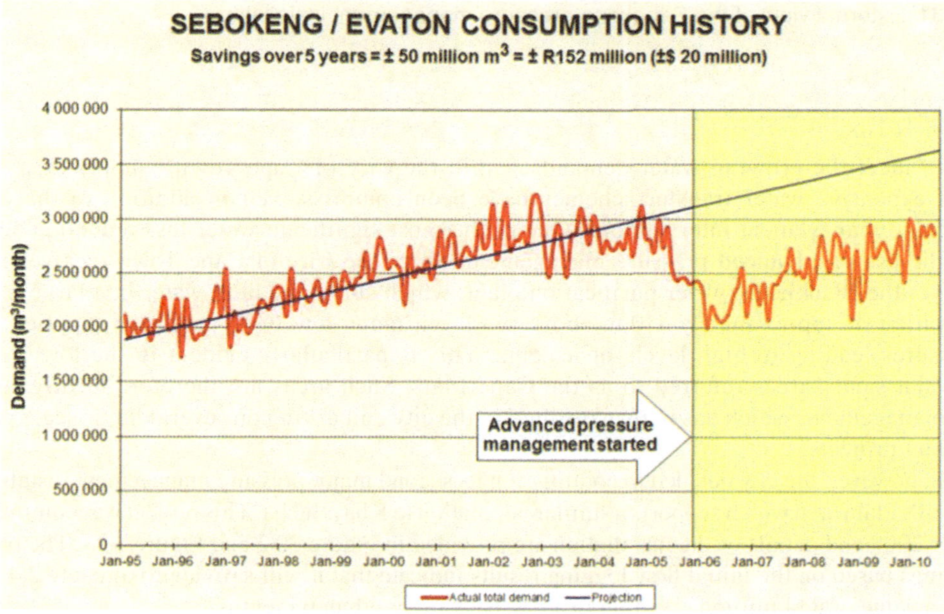

Figure 3.30 Historical water consumption in Sebokeng and Evaton areas for a 13-year period.

The completed Sebokeng/Evaton pressure management installation is shown in Figure 3.29 and Figure 3.31.

Figure 3.31 External view of the Sebokeng pressure management installation.

3.1.11 Mitchell's Plain: City of Cape Town – November 2008

In order to meet the growing water demands within the City of Cape Town's supply area, several large and expensive water transfer schemes have been commissioned in addition to the ongoing water demand management interventions. One of the most significant water loss reduction activities involves the use of advanced pressure management. Since the City of Cape Town receives most of its water via the Blackheath water purification plant, which stores the bulk water in a large reservoir at an elevation of approximately 110 m above sea level, many low lying areas are supplied at very high pressures leading to high levels of leakage. This is particularly evident in the low lying and relatively flat sandy areas referred to as the Cape Flats. Such areas are therefore ideally suited to pressure management, which has in turn resulted in the city embarking on several large-scale pressure management projects.

In 2008, however, the city decided to commission its second major pressure management installation in the Mitchell's Plain area which supports a similar population to Khayelitsha. This project was commissioned in October 2008 and details of the installation are provided in Figure 3.32 and Figure 3.33. The projected water savings based on the initial flow logging results indicate that it will save approximately 2.4 million m^3/a with a value of R14 million resulting in a pay-back of less than 6 months.

The initial savings from the Mitchell's Plain installation are shown in Figure 3.34 (Meyer *et al.*, 2009).

In order to promote the installation and emphasize the importance of water demand management in the community, the installation was painted by local artists in such a manner that it sends a strong message to all passers-by, as can be seen in Figure 3.35.

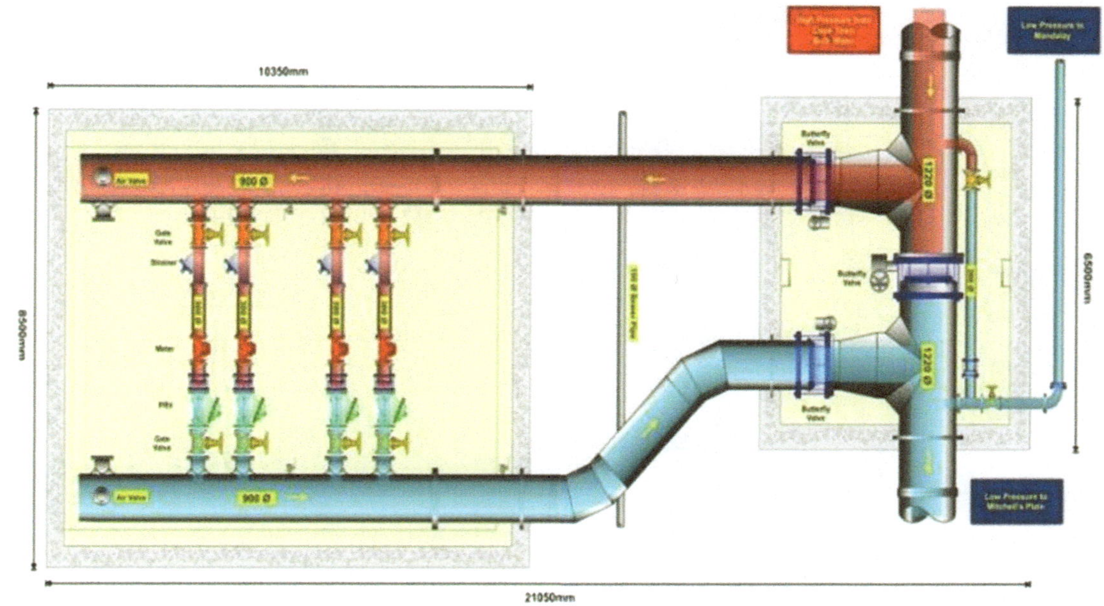

Figure 3.32 Schematic layout of the Mitchell's Plain installation.

Figure 3.33 Mitchell's Plain pressure management installation.

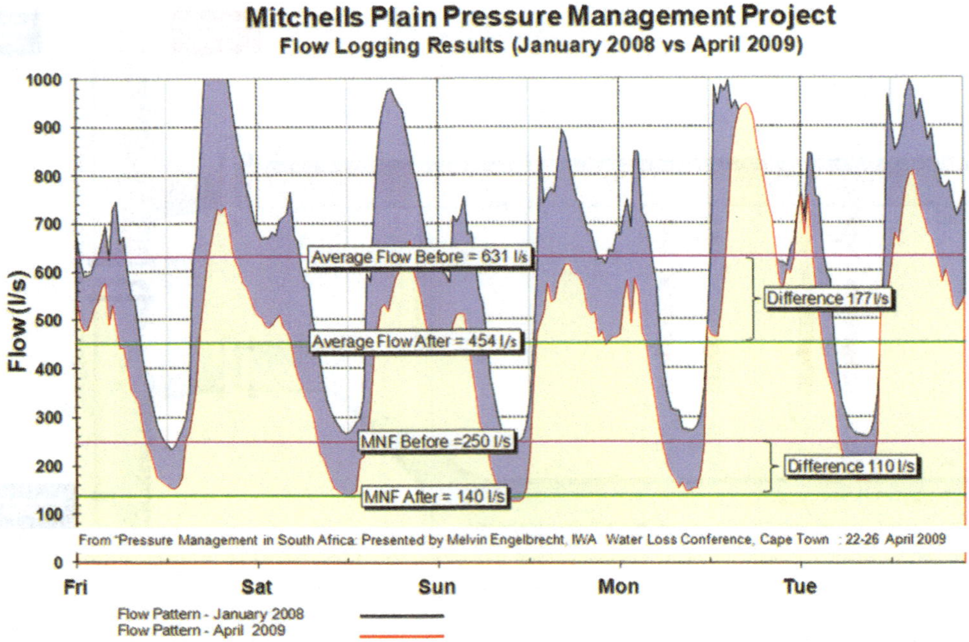

Figure 3.34 Initial savings from the Mitchell's Plain pressure management installation.

Figure 3.35 External view of the Mitchell's Plain installation.

3.1.12 Conclusions

Based on the results from the three case studies, it is clear that pressure management is highly effective in many parts of South Africa and that it can be implemented successfully on a large scale in certain areas. While it must always be remembered that pressure management is normally only the first phase of a larger water demand management strategy, it can often provide very significant savings in a short period of time. Pressure management also has many secondary benefits that are often overlooked, including extending the useful life of the water reticulation system as well as reducing new bursts. Such a benefits can often far outweigh the initial benefits as calculated from the water savings but will only become apparent many years down the line.

South Africa was one of the first countries to recognize the benefits that can be derived from advanced pressure control and currently lays claim to three of the largest advanced pressure control installations in the world. The Khayelitsha and Sebokeng installations in particular have received considerable recognition from various water utilities and funding agencies throughout the world and have often been used to highlight 'world best practice' in the field of water conservation in action.

Finally, it must always be remembered that the most difficult and time-consuming part of any large pressure management installation is usually related to the sectorizing and operation of the pressure management zone. If the basic groundwork has not been completed properly, no pressure management installation will function properly. Unfortunately this is often overlooked, as is the regular maintenance needed to ensure that pressure zones have not been compromised by the unauthorized opening of boundary valves or the closure of internal reticulation isolating valves. These issues are critical and many well designed pressure management installations are not functioning due to problems within the zone. No amount of electronic equipment or sophisticated software will overcome the efforts of a local plumber focused on opening boundary valves. The best option is to introduce some form of monitoring system that will pick up such problems soon after they occur. The monitoring and maintenance of the zones is a continuous process and should always be part of any overall pressure management strategy.

Area	Water Savings (million m^3/yr)	Cost (R)	Savings @R6.20/ m^3 (R/ year)
Khayelitsha	9 million m^3/yr	2.7 mill (2001)	R 55 million/yr
Mfuleni	0.4 million m^3/yr	1.5 mill (2007)	R 2.5 million/yr
Gugulethu	1.6 million m^3/yr	1.5 mill (2008)	R 10 million/yr
Mitchells Plain	2.4 million m^3/yr	7.7 mill (2009)	R 15 million/yr
Total	13.4 mill m^3/year	13.4 mill	R83 mill/yr

From 'Pressure Management in South Africa: Presented by Melvin Engelbrecht, IWA Water Loss Conference, Cape Town : 22-26 April 2009

Figure 3.36 Savings from selected pressure management projects in Cape Town.

The City of Cape Town is one of the most progressive municipalities in South Africa with respect to pressure management and together with the City of Johannesburg, they have been implementing various forms of advanced pressure control since 2000. A summary of their key projects was presented by Meyer *et al.* at the *IWA Water Losses Conference*, held in Cape Town in 2009 and is provided in Figure 3.36. The figures provided refer to only four of the numerous pressure management projects that have been implemented in the Greater Cape Town area but they clearly highlight the significance of the savings that are being achieved by such WDM interventions. Although some of the figures cannot be directly compared with each other without taking inflation and interest into account, it is very clear that the 'pay-back' for the pressure management activities is well under a year and in any municipality, such interventions will be cost effective.

REFERENCES

Ledochowski W. (1956). An analytical method of locating leaks in pressure pipelines. *Transactions of the South African Institution of Civil Engineers*, 341–344, December.

Mckenzie R. (2010). Pressure management pay-back: success for the Sebokeng & Evaton public private partnership. *Water 21*, August, 41–42.

Mckenzie R. and Wegelin W. A. (2003). Leakage reduction through pressure management in South Africa. Paper presented at the *2003 IWA Water Loss Conference,* Lemesos, Cyprus.

Mckenzie R. S. and Wegelin W. (2005). Sebokeng/Evaton pressure leakage reduction: public private partnership. *Proceedings from the International Water Association Specialist Conference: Leakage 2005*, Halifax, Nova Scotia, Canada, pp. 382–391, 12–14 September.

Mckenzie R. and Wegelin W. A. (2009). The scope for pressure management in South Africa. Paper presented at the 2009 *IWA Water Loss Conference*, Cape Town, 26–30 April.

Mckenzie R. and Wegelin W. (2010). PPP to deliver advanced pressure management: success in South Africa. *International Water Association, Water Utility Management International*, **5**(3), September.

McKenzie R., Mostert H. and Wegelin W. (2003). Leakage reduction through pressure management in Khayelitsha, Western Cape: South Africa. Paper presented at the *Australian Water Association Annual Conference*, Perth, 7–10 April.

Meyer N., Wright D. and Engelbrecht M. (2009). Large scale pressure management implementation in the City of Cape Town, *IWA Water Loss Conference*, April.

Chapter 4

The repair or replace dilemma

4.1 DEFINITIONS OF REPAIR AND REPLACE

A repair to a water pipe covers any activity which reduces the unintended loss of water from a pipe due to the failure of that pipe. It may include the placing of a repair clamp around a pipe, or the cutting out of the failed section of pipe or the failed fitting to be replaced by a new piece of pipe or fitting. It may also include the remaking of a joint, such as with run lead jointing. It may even include ad hoc repairs such as the use of wooden plugs or tape wrapping.

Replacement may include relining a pipe with a structural or semi-structural liner, replacing a pipe whilst still leaving parts of the old pipe *in situ*, for example, as in pipe bursting, or replacing the pipe in its entirety, the new pipe either being installed via open cut methods or trenchless options such as horizontal direct drilling.

4.2 THE DURABILITY OF A NETWORK

Managing a distribution network is an ongoing activity. Once water mains are installed, they need to be managed and replaced on an ongoing basis in order to maintain a piped water service. If replacement is not carried out at the appropriate time, the rate of mains failure and leakage will gradually increase. It is possible that the pipes may become so fragile that the disruption caused by carrying out a repair in one location may cause a failure elsewhere, such that it becomes almost impossible to reduce the level of losses. The failure rate increases dramatically at this point and both water losses and costs increase substantially (Figure 4.1, Parker, 2005). It is obviously important to avoid a distribution network deteriorating to this level.

Although replacing water pipes is expensive and should never be considered without looking at the costs and benefits, it is likely that all pipes will reach the point when the cost of repairs and the losses will substantially exceed the cost of replacement. This was shown to be the case for some water companies in the UK, where a substantial mains replacement project is being carried out. The pipes in this case are extremely old, some have been in the ground almost 200 years. However, age does not necessarily indicate that pipes need replacing and other factors need to be considered, as is discussed later in this chapter.

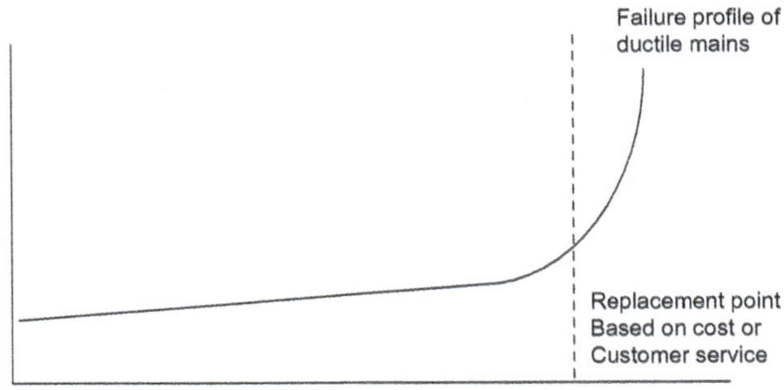

Figure 4.1 (*Source*: Parker).

4.3 DETERIORATION OF A DISTRIBUTION SYSTEM

Different parts of the network may deteriorate at different rates owing to local factors such as ground conditions, pressures, quality and date of installation. The mechanism by which pipes deteriorate and finally fail can be described in three stages.

In the first stage an initial, often minor, defect allows further deterioration to develop. Such a defect may be minor material manufacturing defects, or minor material damage caused during installation. The second step, the deterioration process itself, depends on the material type, for example, corrosion of ferrous pipes, softening of asbestos cement or crack propagation of plastic pipes. Finally the failure itself is often triggered by some random event, which may not be related to the cause of the deterioration.

Such a random event may be

- Ground movement due to changes in soil moisture, frost load or frost heave
- Weather effects (which may effect soil or water temperature or soil moisture)
- Increasing traffic loads
- Successive internal pressure (water hammer, changes in pressure)
- Failure of adjacent utilities.

These stages may affect even quite new pipes.

4.4 FACTORS AFFECTING THE RISK OF FAILURE OF WATER DISTRIBUTION PIPES

The following is a list of presumed risk-of-failure factors, as reported in past research studies:

- Pipe age – the older a pipe is, the more likely it has reached a high grade of deterioration.
- Pipe diameter – large-diameter pipes will operate under different stress patterns, such that their pattern of failure is likely to be different from smaller diameter pipes.
- Pipe material – a pipe's material affects the pipe's resistance to loads and deterioration.
- Pipe length – a pipe's length increases the chances of breakage.
- Number of previously observed breaks – this indicator is important in relation to the frequency of fractures and a pipe's behaviour diachronically, as well as indicative of how susceptible the pipe is for future fractures. The higher the number is, the more susceptible the pipe becomes.
- Pipe pressure – the higher the internal pressure is, or the more frequent the fluctuations in it are, the higher the chances are for failure. Thus, it is important that the network's operating pressure is kept as uniform as possible.

- External temperature – the external temperature appears to contribute to the frequency of fractures (historically increasing during the summer months due to changes in soil moisture leading to soil movement).
- Surroundings – the region (surrounding space) in which a pipe is located influences its chance of failure. For example, since pipes in industrial neighbourhoods are daily subjected to adverse loading conditions, they are more likely to break.
- Traffic loads – pipes in the proximity of heavily trafficked areas are more susceptible to failure, especially if laying depth is low.
- Construction loads – pipes located in areas under construction are more susceptible to failure.
- Soil type – the type of soil in which a pipe is buried influences its behaviour and risk of failure. For example, dry soils are more likely to cause 'beam effects' on pipes owing to soil movement. Conversely, wet soils cause erosion to metallic pipes.

Kleiner and Rajani (Kleiner & Rajani, 2001) summarised these factors in the diagram presented in Figure 4.2.

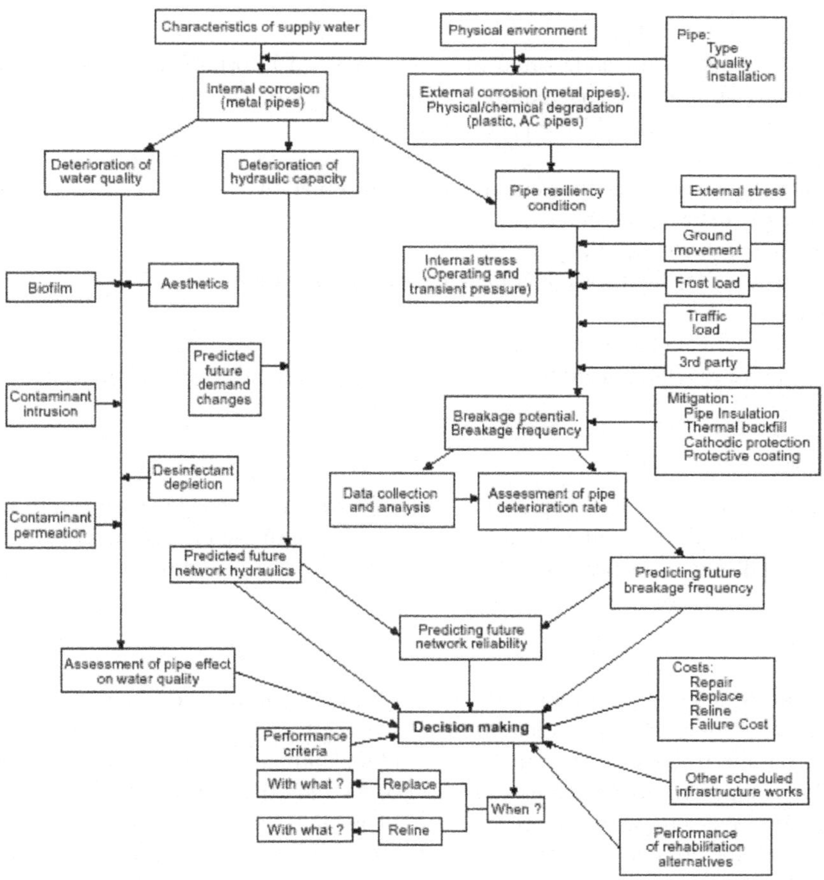

Figure 4.2 Factors and mechanisms feeding into the decision-making process: holistic view (*Source*: Kleiner & Rajani, 2001).

4.5 DEFINITION OF MAINS FAILURE

Failures may be very small and difficult to detect or substantial leading to a major escape of water, reductions in pressure and loss of supply to customers. Generally the smaller escapes are called 'leaks' and may not be detected for some time. The larger escapes of water are often identified very quickly, may lead to substantial columns of water due to the pressure within the pipe, and may be referred to bursts. The generic term used throughout these notes is 'failure' which will cover all escapes of water.

4.6 THE IMPORTANCE OF INFORMATION

In order to make informed decisions about the distribution network it is important to have as much information about the network as possible. Although the pipes are buried, it is possible to develop a good idea of the current condition of a water pipe and also to estimate how that will change over time.

4.6.1 Problems with information collection

Most distribution engineers will understand the need for good information. However, the information available often falls short of what is required with a number of shortcomings, even in utilities with sophisticated information systems. These problems include:

- Lack of knowledge regarding the veracity of past information
- Inability to link different databases
- Lack of audit to confirm accuracy
- Lack of motivation for field staff to collect accurate information
- Gaps in data
- Too much data being entered with default values
- Trying to capture too much information and ending up with only a partial data base.

4.6.2 The 'Garbage In, Garbage Out' (GIGO) principal

With all asset information, the GIGO principal applies – if you put garbage in then you'll get garbage out. This means that time spent developing initiatives to train field staff about why information is needed and what is needed is well spent as is time developing processes to audit information.

4.6.3 Minimum requirements

The very minimum which every utility should aim for is a plan of all water pipes placed against a map or aerial photograph background. Each length of pipe as well as all valves and other fittings should be identified with a numbering system such that they can be identified and the status of the valve recorded. This is particularly important for boundary valve to DMAs and other zoning. Additional data about each pipe is also important, particularly the diameter, material and the year laid, which is as accurate as possible. It is better to describe the possible decade of laying than providing no information. Additional information is necessary to describe the accuracy of data in this context.

In order to be able to predict failure rates or failure occurrence time as well as the risk associated with failures, the following information should be collected and stored in such a way that it can be cross-referenced (e.g., link pipe IDs and other attributes with pipe failures) and geo-referenced (to be able to identify them on a GIS system):

- Pipe: diameter, vintage, material, corrosion protection (inner, outer) joint-type, number of connections, history (original, renovated), and status (operating, decommissioned)
- Failure data (date, reason, failure mode, repair type, cost [or possibility to link to other systems like SAP], etc.)
- Operational data – pressure (max, min, average, etc.)
- Soil characteristics (e.g., corrosivity)
- Traffic data
- Water quality data.

4.6.4 Developing information systems

This basic system can be developed and linked in to other systems such as customer databases, providing property numbers and consumption figures, leakage management systems which record night flows and calculate leakage in each DMA and network modelling systems which can help provide answers about how the network is currently performing and will perform in the future.

Ensuring that data can be transferred between systems is important to eliminate double entry of data and to facilitate the development of information systems which will provide the answers as to whether a pipe should continue to be repaired or whether it should be completely replaced. Some GIS systems for instance produce data in formats which are not easy to transfer to other systems.

The process of lining databases and the GIS system may require detailed planning and may take several years to implement (Hladej, 2007)

4.7 DISTRIBUTION PERFORMANCE MEASURES

4.7.1 The importance of measuring performance

It is important to collect a number of measures for a distribution system. The choice of what to measure, where to measure it, how to collect the data and at what frequency, should be made by the utility. However, it is important that indicators for the service received by customers, the condition of the pipe network itself and the cost of running the network are collected in order to be able to assess whether replacement is a cost effective option.

4.7.2 Performance measures for distribution

There are a number of performance measures which can be used for distribution systems. The Natural Rate of Rise measures the rate at which physical losses (leaks) occur and grow and can be used to analyse both the need and the cost benefit of carrying out mains renewal (Grimshaw, 2006, 2009). The quantity of water lost will also give an indication of the state of the network, but care should be taken as it is also heavily dependent on non-revenue water management policies. Interruptions to supply due to major failures also provide an indication, but again care must be exercised as these may be dependent on pressures in the system.

The *IWA Manual of Best Practice* also offers guidelines on what performance measures to collect with defined performance indicators (PIs), such as water loss PIs (e.g., ILI); failure PIs (mains, valves and service connections – further defined whether bursts or detected failures), leakage control (%/y), active leakage control repairs, rehabilitation PIs.

4.7.3 CARE W

The European project CARE W developed guidelines for the continuous observation of the performance of the network and the utility and provides definitions and evaluation of asset performance indicators (concerning pipe failures, water losses, customer complaints etc.) that can be calculated on the network, zone or even the pipe level (Allegre *et al.*, 2005).

4.7.4 Cost information

The collection of costs is an important part of monitoring performance. Without this, an investment plan cannot be based on a sound economic argument. Accounting practices vary from country to country and even between utilities in the same country. However, there does need to be some differentiation between operating costs – those costs which are incurred as a result of day to day operations – and capital costs, which are generally major items of expenditure that expand the asset base or extend the life of the assets.

Possible costs to consider as operating costs

- repairs
- valve operations
- flushing
- leakage location
- informing customers
- compensation for flood damage
- handling customer complaints
- increase in water treatment and pumping costs because of water losses, incrustation
- Socioeconomic costs caused by supply interruptions, traffic interruptions, noise and dust of pipe repair construction sites.

Possible costs to consider as capital costs

- materials used to replace mains
- labour used to replace mains
- purchase of land if required
- procurement costs
- cost of borrowing.

4.7.5 Environmental factors

The failure rate of a distribution network may vary with external factors. Some of these include temperature, soil moisture, road loading and useage, road surface and adjacent activity.

The effect of different climate factors and particularly changes in ground temperature are still being researched to understand the effects better. The changes may occur in a variety of ways, such as low or high ambient temperature, as well as the temperature of rainfall. The above phenomena create a dynamic and complex system which affects underground pipes laid at different depths.

Some of the phenomena that ground temperature variation creates or increases the significance of are:

- Damp or wet ground that dries out, creating stress on the pipeline and in more severe cases ground movements creating bursts.
- Expansion and contraction of plastic and metallic pipes.
- If the ground temperature reaches the below freezing point, the water inside the pipes expands, creating stress on the pipes, and the ground around the pipes also freezes creating even more stress due to the additional frost load.
- When the ground temperature rises from freezing point and the ground begins to thaw, ground movement creates further stress.

The above are some of the factors that affect the pipelines due to ground temperature variations. Several failure prediction models already incorporate a number of climatic indicators and allow a better prediction of failure frequency resulting from severe climate conditions (e.g., Kleiner *et al.,* 2010). More research on this subject by specialists, such as geotechnical engineers, could help us understand these problems better.

4.7.6 Configuring the data

Modern computing provides excellent facilities to collect the data. However, financial systems are not always configured to facilitate the abstraction of data to be used for cost benefit studies. Ideally, costs should be able to be allocated to a specific pipe or zone. This may require consideration of financial coding systems and should be taken in to account for any new systems.

The IWA recommends the use of DMAs to monitor, localise and manage water losses. This unit is also a very useful tool when collecting performance data for water distribution networks. If the data collected has an extra field in which the DMA reference can be entered then the data can be analysed on a DMA by DMA basis. This is particularly useful for a large network as DMAs can be prioritized in order of performance, as shown later in this manual.

4.8 A MAINS FAILURE DATABASE

4.8.1 Why mains failure information is needed

One of the most important pieces of information is which pipes have failed when and if at all possible, how they failed. Whilst this information is often collected, it is not always collected in a format which makes it easy to interrogate. Often the information was originally collected in order to manage the workforce or pay a contractor, and it may not be recognized as valuable in itself.

It is important to build this information up at an organizational level so that patterns in deterioration can be identified and sound cases can be made for capital investment where it is found to be needed. Rates of failure by material, age and material on a pipe by pipe or zonal basis should be able to be calculated.

4.8.2 National failure databases

In the UK, a national standard has been agreed with three different levels for failure information, basic, preferable and desirable (MacKellar & Pearson, 2003). Basic information includes the date, location and the type of activity, diameter and material of main. Collecting this information has proved invaluable and all water companies use the information to identify the pipes which have the highest risk of failure in the future. The database can be further enhanced by extending the information to cover, for example, soil sample results, whether this was a repeat visit, the repair technique used and the jointing technique used for the pipe (Figure 4.3).

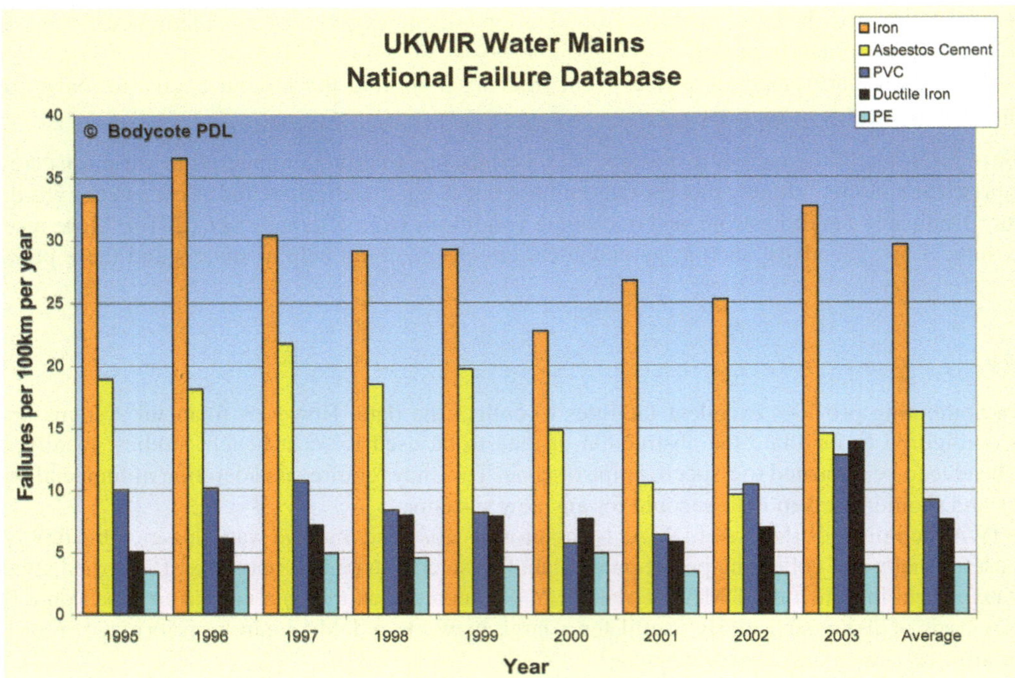

Figure 4.3 Early outputs from the UKWIR National Mains Failure Database (*Source*: UKWIR).

All the UK water utilities also submit data to a national database, which means that there is a large volume of failure data which can be investigated to identify the principal factors which affect failure. Combining failure data in this way can be beneficial as it can provide a much larger sample from which different trends can be ascertained. A number of other countries have also developed or are in the process of developing national failure databases.

A number of these national databases now provide a web interface so that data can be uploaded and accessed through the web. The data is also being analysed to identify certain trends, which is helping water utilities to identify the key factors which affect water main deterioration.

4.9 STARTING FROM SCRATCH

Not all organizations have a robust asset database or performance information system. However, much of the information they will need can be collected and may already be available through, for instance, their leakage or repairs system.

4.9.1 Compiling a mains asset database

Good records of the mains system in the ground is the first step. The availability of this information may be dependent on earlier operational policies as past record drawings and plans may have been lost, destroyed or never prepared. If there are gaps in the information, it is essential to mobilize all employees and contractors working on the distribution network to collect information at every opportunity, for instance when repairs are being carried out. Other utilities may also be able to help, providing information if they have a cause to carry out excavation work and expose the water main. Whenever a water main is exposed, there is an opportunity to check the material, diameter, age, type of soil surrounding it and the exact location. This will help improve the database.

4.9.2 Capturing repair information

When repairs are made, it is important to collect as much information about the repair as possible. How the need for a repair was identified should be captured as this may indicate whether the leak developed rapidly or slowly over time. The time of identification may not be the exact time when the failure occurred, but when dealing with pipe lifetimes of decades, the inaccuracies are generally not important. Information about the failure mode should also be collected and simple charts or sample photographs to explain how to recognize the different types of failure, for exmple, ring fracture, failure of a ferrule, and so on may help field staff. Pipe cut-outs should be collected at every opportunity – when connections are made, as well as from repairs. Soil samples collected at the same time can help provide information about factors affecting the pipe condition. Both pipe and soil sample collection can be encouraged by providing plastic bags with labels ready attached, which just need details of the date and location of the work.

4.9.3 Collecting additional information

Although a substantial amount of information can be obtained through monitoring failures, some additional information about the current condition of the pipe network may be useful. One approach is a proactive sampling programme, rather than just taking opportunistic samples. These can be very useful to identify for instance the level of corrosion for ferrous mains in order to help populate statistical models to determine the rate of deterioration.

There are, in addition, a number of non-destructive tests available. These include internal cameras and sonds, which are now available for pipe diameters down to those of a normal distribution network, although the majority are designed for larger diameters. Access points may be required and the cost of these should be added to the cost of the specialist equipment and if required expertise to interpret the results. However, most systems will now work on live pipework with facilities for ensuring the hygiene of the system is maintained.

4.9.4 Organising the information

Again, if DMAs have been established they can provide a useful unit for comparison. Repair levels, leakage levels, rate of rise and, if available on a DMA basis, costs per DMA can provide a simple basis for prioritization of areas for renovation and replacement. Other factors such as water quality problems can also be taken into account (Figure 4.4, Parker, 2001).

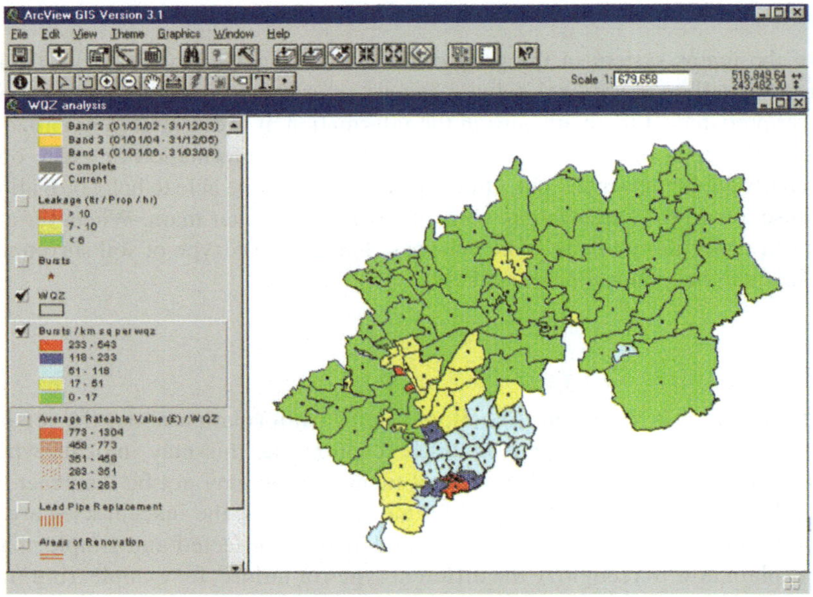

Figure 4.4 Burst information displayed on a zonal basis across a utility's complete area (*Source*: Parker, 2001).

4.10 ANALYSING THE INFORMATION

4.10.1 The simplest approach

Although some utilities use sophisticated methods to analyse the data, such as described later, a simple spreadsheet system such as Excel combined with a basic GIS system on which failures are marked will provide a lot of valuable information. From this it will be possible to identify some trends in failures. Which areas have the highest failure rate? Do failures occur at certain times or after certain weather? Do pipes made of certain materials or of a certain age fail more frequently?

4.10.2 Use of a GIS system

It helps to be able to analyse data both in groups of different categories such as pipeline material or age and also analyse it geographically. Many GIS systems have excellent database facilities within the package and thus the information can be analysed and viewed on a geographical basis, which may flag up localized problems, due to, for instance, local soil conditions (Figure 4.5, Parker, 2008).

Examples of a GIS-based decision support system used in Cyprus are shown in Figures 4.5 and 4.6. At first, historical data on previously observed breaks (NOPB) are lumped at a street level and then mapped to a GIS map of the pipe network, colour-coded to indicate the variable degrees of their inherent risk of failure (Figure 4.5). The Water Boards can therefore easily and holistically review the status of their network in terms of where and how often pipes break, as well as the computed risk-of-failure for each segment of the pipe network. Even though the eventual goal is to calculate risk-of-failure metrics at the pipe level and not the street level (in other words an individual forecast for each pipe segment) it was deemed redundant and over-complicating at this early stage of the research and thus overlooked in favor of metrics for each street segment.

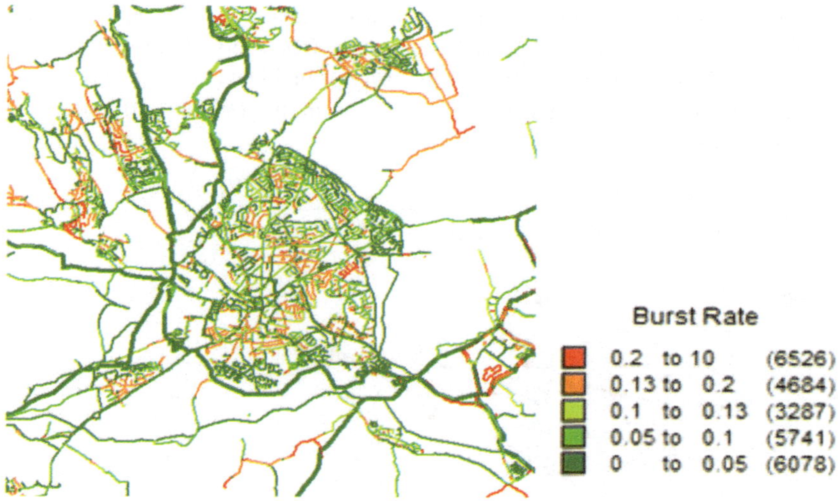

Figure 4.5 Example of geographical output using a GIS system (*Source*: Parker, 2008).

Figure 4.6 Colour-coded GIS mapping of risk-of-failure (*Source*: Christodoulou *et al.*, 2007).

The data is also categorized by the type of pipe for which burst incidents are reported, and graphed in histograms also at the street level (Figure 4.6). For example, for two case-study streets (Valtetsiou and Kosti Palama), the histograms indicate the majority of incidents are AC and MDPE pipes respectively.

Should one break the data down further, the data can be categorized by pipe type (mains or house connections) or incident type (connection failure, corrosion, tree roots, deterioration due to aging), as shown in Figure 4.7. The histograms show that the majority of incidents are due to tree roots (Valtetsiou) and aging (Kosti Palama), with most incidents observed on mains (Valtetsiou) and house connections (Kosti Palama) respectively.

Figure 4.7 Histograms of incident categorization (*Source*: Christodoulou *et al.,* 2007).

4.11 DEALING WITH HIGH BURST RATES

4.11.1 How to decide if the failure rate is too high?

Whatever method is used to analyse the failure data, some areas or pipes may be identified as having a particularly high failure rate or risk of failure. Just what constitutes high depends on the utility's policy, for example, what level of customer interruptions is acceptable? What repair cost is tolerated? There may also be regulatory limits imposed on leakage and this may affect the decision. The analysis may consider failure numbers, water losses, rate of increase leakage (rate of rise) or a combination of all of these. Whilst it may well be that these pipes need replacing, it is worth checking a number of other issues first.

4.11.2 Pressure control to reduce failure rates

The relationship between pressure and burst rate has been proven for some time (Thornton & Lambert, 2005). Managing the pressure in a zone may reduce burst levels to acceptable levels without any replacement. Even where average pressures are not excessive, pressure variations can lead to higher burst rates and so installing pressure control can still reduce burst rates and prove cost effective (Pearson *et al.*, 2005). Work in Sao Paulo, Brazil, Selangor, Malaysia and Ho Chi Minh City, Vietnam showed benefits can be obtained even where starting pressures are low (Thornton *et al.*, 2005). A more recent example in Lucca, Italy showed that substantial improvements in both burst rates and leakage can be obtained by managing pressures and reducing pressure fluctuation, eliminating the need for immediate mains replacement (Rogers, 2008). Pressure control has other benefits and may well cost considerably less than a mains renewal project. Therefore it is recommended that pressure control is considered before any extensive mains renewal is planned.

Intermittent supplies may also shock pipes on a regular basis such that failure may occur more frequently. Restoring 24-hour supply can reduce the occurrence of failures and generally major replacement schemes should be avoided until 24-hour supplies have been established, the reason being the likelihood of elevated failure rates.

4.11.3 Increase of failure rates with increased pressure

Conversely, where leakage activity or other operational changes lead to an increase in pressure, it is possible that burst rates may increase, as was found in the Bahamas New Providence district (Fanner, 2007).

Pressure control will not eliminate the deterioration of the main and it will be necessary to replace all mains eventually. However it will extend the life of pipes and allow major capital schemes to be delayed, which can be valuable in itself. This should be taken into account in designing any pressure management system. For further information on how to assess and design pressure management systems, refer to the IWA publications on pressure management.

4.11.4 Identifying mains replacement projects

If pressures are actively controlled and kept within reasonable operating limits and failure rates and leakage is still unacceptably high then a mains replacement project will need to be considered.

Any identified schemes need to be prioritized. This may be done just on unit burst rates, for example, bursts per kilometre per year. It may be possible to develop software packages which identify pipe lengths which have high failure rates. As explained earlier, it may be easier to carry out a coarse prioritization exercise by plotting burst rates on a zone by zone basis. The zones with high burst rates can then be examined in more detail to identify the pipe lengths which require replacing.

The exact length of main to be replaced needs to be considered carefully. A major part of any mains replacement project is the cost of 'setting up', that is, of moving plant and equipment to the location, establishing a site office if required, liaising with other authorities, and so on. It is worth considering other lengths of pipe in the vicinity, which may not have such a high failure rate but which become economic to replace whilst the other work is being done. Other factors such as known road resurfacing projects or other local initiatives may also drive decisions.

GIS based analysis to define the length of the pipe segments, which are affected by pipe failures will provide an economic analysis to provide the optimal time of rehabilitation using failure/operating costs prediction as well as a priority ranking within the area further.

A further issue is whether service pipes should be renewed as well. This question can be answered by using an analysis of failures on the different components of the system: mains/service pipes/valves.

4.12 COST BENEFIT STUDIES

4.12.1 The need for cost benefit studies

In some cases, a renewal scheme may be driven by customer service, for example, if repeated failures on one section of pipe lead to repeated failures in supply for a group of customers. However, even in these cases, a cost benefit study should be carried to understand how operational costs may reduce and possibly to further prioritize projects.

4.12.2 Calculating direct costs

Direct costs to be taken into account for a mains replacement project include design costs, the cost of any contractor hired or the time taken by direct labour, the cost of materials and the cost of ancillaries, such as any traffic controls, purchase of land or negotiation of way-leaves, public consultation, and so on.

It is important to have good estimates for the potential costs of replacement. Actual costs for replacement schemes should be recorded along with details of the length, diameter, pipe material (both original and replacement), method of replacement, location, surface and soil characteristics. Over time, this will allow comparative costs to be developed, which can be used for future estimates.

Pipes do not always need to be laid in open trenches and installation using 'trenchless' methods can be cheaper in some cases, particularly where traffic and other surface damage and disruption is a major factor. Trenchless techniques have generally been found to have a lower impact on the environment and carbon footprint (Burtwell *et al.*, 2005).

4.12.3 Potential savings

This should be compared to the savings accrued from reduced leakage location activities and repairs as well as reduced water losses. If there is an overall shortage of water and the water saved can be sold, then there is some justification in using the retail price of water. If the production cost of the water is used, then it is important that the pumping cost to the zone in question is considered. If leakage in the distribution network is high then losses elsewhere in the system should be factored in. Although the new pipe will start to deteriorate and there will be gradual rate of rise in leakage, this is likely to be much lower than for the old pipe removed. However there is little information about the rate of rise in new pipe systems and an estimate will need to be made (Grimshaw, 2009).

4.12.4 Comparing capital and operational costs

Capital and operational costs should be compared using a net present value approach. This will require a discounting rate to be applied which should take likely current and future economic conditions into account.

The economic optimal time of rehabilitation can be calculated by using the discounted cash method to compare future costs on old and new pipes, thus identifying future savings. However, the sensitivity and quality of the result of the economic optimal time of rehabilitation should be investigated in more detail to asses, for example, which costs have to be taken into account for the calculation depending on the system (e.g., are there high or low pumping or water treatment costs….)

4.12.5 Indirect costs and triple bottom line approaches

There is also some discussion as to what costs should be included for a cost benefit study. Whilst direct costs, i.e. those costs incurred by the water utility are generally taken into account in any cost benefit study, there are other costs which may not be considered. Some water utilities now monitor performance measures other than financial measures and in particular sustainability measures such as carbon emissions may be considered important. This approach is sometimes called 'triple bottom line' accounting where in addition to financial measures, environmental and societal measures are also taken into account. Both mains repairs and renewals have an impact on the environment and society (Parker, 2006).

The social costs include such impacts as the disruption to local businesses whilst environmental costs include the need to excavate and transport crushed stone for the reinstatement of roads.

In the UK, the approach to costing leakage reduction activities has been captured in the report 'The Environmental and Social Value of Leakage Reduction' (Atkinson *et al.*,1999). Further work to assess the cost of street works has been published in a later UKWIR report (Burtwell *et al.*, 2005).

4.13 PREDICTING FUTURE TRENDS

4.13.1 Why predict the future?

So far, we have considered information about current and historic performance. However, capital investment schemes often take several years to plan and a view of capital investment needs in the future may be required for planning purposes. In addition, utilities should assess whether the state of the distribution network is being maintained, is improving or less desirably, is deteriorating. Regulators may require water utilities to demonstrate that they are maintaining the condition of their assets.

In these cases it is necessary to develop numerical models to represent the deterioration of the distribution network. This should take into account the different ages and materials of the networks as well as other factors which will affect the life of the pipes, such as ground conditions, weather and ground loading.

4.13.2 Models available

A wide variety of approaches and models exist to identify how pipes will deteriorate and fail in the future. Physical models require a substantial amount of data and therefore for distribution mains the models are frequently statistical. A number of these were reviewed by Kleiner and Rajani (2001). Some of these are outlined below.

The previously mentioned European project CARE W has developed a suite of programmes which helps define both long term strategies for pipeline rehabilitation and also prioritizes annual rehabilitation programmes. (Herz *et al.*, 2005; Le Gauffre & Torterotot, 2005). The Pipeline Asset and Risk Management System (PARMS) was developed by the Commonwealth Scientific and Industrial Research Organisation (CSIRO) in Australia and is used by a number of Australian water utilities (Burn *et al.*, 2004).

In Austria, a rehabilitation planning system (PiReM) was developed and applied to several utilities in Germany and Austria based on a proportional hazard model and on Herz-function (Herz, 1996). The Herz model is considered for survival rates estimation while the proportional hazard model is used for accelerated failure time predictions to support pipe prioritization. (Fuchs-Hanusch *et al.*, 2011; Fuchs-Hanusch & Kasess, 2012).

In Cyprus, a predictive approach was developed to assess mains replacement requirements in Lemesos (Christodoulou *et al.* 2007). This considers the hazard and survival rates as shown in Figure 4.8. The visual analysis illustrated in Figures 4.6 and 4.7 can be coupled with numerical analysis of the hazard rate and survival plots, 'stratified' by different logical groups.

In this example, survival analysis of the pipe incidents grouped by the type of incident (such as pipe deterioration, interference by others, tree roots, corrosion, etc.) revealed interesting patterns in terms of the causes for pipe failure. Looking at the hazard rate over the time (measured in days of presumed pipe age) it is possible to see that the rate with which the hazard for failure increases for pipes experiencing deterioration is faster than the hazard rate for pipes under 'interference by others' (Figure 4.8). It can also be seen in the same plot that the hazard rate for pipes in the vicinity of tree roots accelerates over time (i.e. the slope gets steeper), indicating that pipes in proximity of tree roots should be replaced at shorter time intervals or otherwise risk failure. It can also be considered that in this location tree roots become an issue for pipe failure as time progresses.

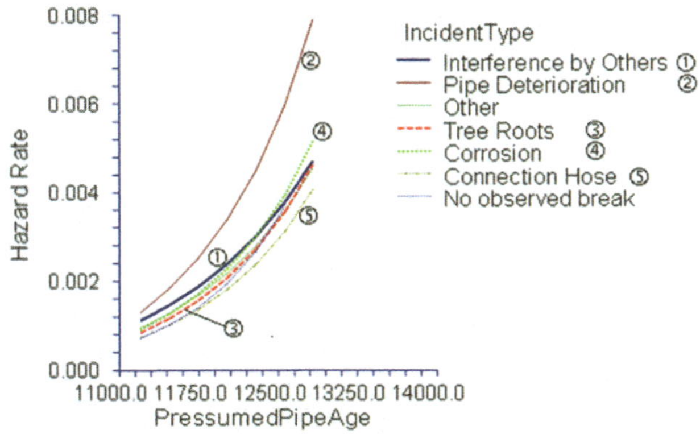

Figure 4.8 Hazard rate plot, by incident type (*Source*: Parker).

In the UK, a risk-based predictive approach to asset maintenance is required by the regulator and so all water companies have adopted some sort of modelling approach. Some consultancies such as Tynemarch and SEAMS in the UK have developed systems to meet the regulatory requirements.

4.14 AVOIDING LEAKAGE IN NEW SYSTEMS

An important part of managing leakage levels in the future is to ensure that the quality of new installations, whether for new development or replacing existing pipes, is of the highest possible quality to ensure that new leaks are not being introduced into the system unnecessarily.

If a system is being replaced, it is important that the problems of the old system really are eliminated and the new system is leak free. A variety of materials may be used for new water mains; uPVC, polyethylene and ferrous (ductile or steel) mains are the most commonly used today. These can all be extremely robust systems with minimal leakage for many years if the appropriate material of adequate quality is installed properly.

4.14.1 Choice of material

It is important that conditions are taken into account when choosing the pipe material. Method of installing, traffic loading, ground conditions, such as aggressive soil and the stability of the soil, are all factors which should be considered. In addition, it is worth considering the level of expertise available for installing the pipes, as well as the quality of manufacture.

4.14.2 Contract requirements

Contract conditions may have an impact on the quality of installation. Details of the depth, bedding and surround for the pipes must be included and the exact line of the main detailed. The mains should be tested before they are commissioned with an appropriate test for the material, and it is essential to ensure that all services have been transferred. Adequate maintenance periods should be allowed to ensure that problems which may not initially be apparent are dealt with. Supervision and audit of the work is important and standard audit sheets can be developed to monitor the work.

4.15 FURTHER RESEARCH

There are still many questions which need to be answered. For instance, if information on distribution networks does not exist, such as their location or material – as is the case in many locations, how best can water distribution networks be mapped and recorded? Can it be done without extensive trial holing?

Work is still ongoing to better understand how the level of leakage or the rate of rise varies over time and further work will help to refine predictive models as we understand more about how pipes deteriorate. The life span of PE pipes has still to be tested in reality, although laboratory tests are promising and there are no signs of the early pipes failing due to material deterioration. There is limited information about the level of leakage in new systems and which of the new materials and methods of jointing is best for minimizing leakage in different situations. It would also help to identify whether some design approaches are better than others.

REFERENCES

Atkinson J., Baker B., Moran D. and Williamson B. (1999). The environmental and social value of leakage reduction, *UK Water Industry Research*, London, UK.

Allegre H., Baptista J. M. and Coelho S. (2005). Performance indicators for water network rehabilitation, computer aided rehabilitation for water networks. In: S. Sargrov (ed.), IWA Publishing, London, UK.

Burn S., Ambrose M., Moglia M. and Tjandraatmadja G. (2004) An Approach to Strategic Management for Urban Water Infrastructure, LEA, Australia.

Burtwell M. H., Evans M. and McMahon W. (2005). The real costs of street works to the utility industry and society. *UK Water Industry Research*, London, UK.

Christodoulou S., Charalambous C. and Adamou A. (2007). Managing the "Repair or Replace" Dilemma on Water Leakages, IWA, London, UK.

Fanner P. (2007). Pressure management works……..and doesn't, *Water Loss 2007*, IWA, London, UK.

Fuchs-Hanusch D. and Kasess D. (2012). Maintenance management – risk- and condition-oriented planning for Vienna's water pipe network, *Proceedings of IWA Water Loss Conference*, Manila.

Fuchs-Hanusch D., Kornberger B., Friedl F. and Scheucher R. (2011). Whole of life cost calculations for water supply pipes, *4th Conference Proceedings of Leading Edge Conference on Strategic Asset Management*, International Water Association IWA, Mülheim an der Ruhr, Germany.

Grimshaw D. (2006). Leakage-driven mains renewal, *Proceedings of Joint ICE/CIWEM Conference: Infrastructure Asset Maintenance and Management*, Birmingham, UK.

Grimshaw D. (2009). NRR – The key to leakage driven mains renewal, *Water Loss 2009*, IWA Publishing, London, UK.

Herz R. K. (1996). Ageing process and rehabilitation needs of drinking water distribution networks. *J. Water SRT – Aqua*, **45**(5), 221–231.

Herz R., Baur R. and Kropp I. (2005). Strategic planning and investment. In: Computer Aided Rehabilitation for Water Networks, S. Sargrov (ed.), IWA Publishing, London, UK.

Hladej M. (2007). City of Vienna – Network Information System – how to record the condition of water distribution systems. *Conference Proceedings of Water Loss 2007*, Bucharest, Romania, IWA, London, UK.

Kleiner Y. and Rajani B. (2001). Comprehensive review of structural deterioration of water mains: statistical models. *Urban Water* **3**(3), October, 131–150.

Kleiner Y. Nafi A. and Rajani B. (2010). Planning renewal of water mains while considering deterioration, economies of scale and adjacent infrastucture. *Water Science and Technology: Water Supply*, June 10, 867–906.

Le Gauffre P. and Torterotot J. (2005). Decision Support for Annual Rehabilitation. Computer Aided rehabilitation for water networks, S. Sargrov (ed.), IWA Publishing, London, UK.

Le Gauffre, Haldar, Poinard (2007). Assessing the benefits of rehabilitation programmes defined with the CARE W decision support system. *Conference Proceedings of IWA Leading Edge Conference on Strategic Asset Management*, Lisbon, Portugal.

MacKellar S. and Pearson D. (2003). Nationally Agreed Failure Database and Analysis Methodology for Water Mains. UK Water Industry Research, London, UK.

Parker J. (2001). 'An integrated approach to Managing Network Assets' IQPC Infrastructure Asset Management Exchange London 2001.

Parker J. (2005). Leakage and the link to asset management, *Leakage 2005*, IWA, London, UK.

Parker J. (2006). Sustainability and social issues – an international perspective. *Conference Proceedings 'N Dig Down Under*, Brisbane, Australia, UKSTT.

Parker J. (2008). Establishing an Appropriate Mains Replacement Strategy. *Conference Proceedings Water Loss, Telemetry and SCADA in Water Distribution Systems* ADKOM Ohrid, Macedonia.

Pearson D., Fantozzi M., Soares D. Waldron T. (2005). Searching for N_2: How Does Pressure Reduction Reduce Burst Frequency? *Conference Proceedings* for *Leakage 2005*, IWA, London, UK.

Rogers D. (2008). An Integrated approach for water network rehabilitation, *World Water Congress 2008*, IWA, London, UK.

Thornton J. and Llambert A. (2005). Progress in practical prediction of pressure: leakage, pressure: burst frequency and pressure: consumption relationships. *Leakage 2005*, IWA London, UK.

Thornton J., Shaw M., Aguiar M. and Liemberger R. (2005). How low can you go? A practical approach to pressure control in low pressure systems. *Conference Proceedings for Leakage 2005*, IWA London, UK.

Chapter 5

Community awareness and education

5.1 INTRODUCTION

For many years, community awareness and education has often been ignored when dealing with Water Demand Management. It is usually considered as a 'nice to have' which is only implemented if there are some unspent funds looking for a home at the end of the financial year. A variety of arguments have been suggested to explain this limiting mind set particularly amongst technical professionals in the water sector. It often comes down to the fact that it is virtually impossible to measure the tangible benefits resulting from community awareness and associated education activities as well as the general absence of proper monitoring and evaluation techniques. The reality of the situation is that many of the technical interventions will ultimately fail if the communities that they are designed to help are not informed properly through appropriate awareness and education support. In most countries, the consumers demand to be informed and involved in the decision-making process. Too often, worthwhile and well-designed technical WDM interventions eventually fail due to the lack of support from the community. The need for proper consultation, and the creation of awareness for the project under consideration cannot be over emphasized. Communication has thus become an essential component to sustainable development throughout the world in which those who are impacted by any proposed intervention, take ownership of the development and the maintenance thereof. This was never more pertinent than when it comes to water, one of the most basic resources and human needs, which is responsible for current and future prosperity in its varying forms.

This chapter thus aims to address some of the perceptions noted above and discusses some of the challenges pertaining to the implementation of community awareness and education. This chapter will further deal with the varying mediums of communication and provide some advice on what is required for successful community engagement on water conservation based on many years of practical experience.

5.1.1 Importance of community awareness and education

As a point of departure, it is vital to understand why community awareness is necessary specifically in the context of WDM as a key strategy for non-revenue water reduction. In recent years, WDM has taken centre stage in the fight against the ever increasing scarcity of water, which is becoming the single most important issue in many parts of the world. High levels of water losses coupled with low levels of payment for services and poor quality infrastructure are experience in virtually every developing country as well as many developed countries. It is no longer adequate, proper or effective for professionals to develop WDM strategies based on assumptions and scientific calculations in the obscurity of their offices. The current economic climate demands a move towards cost-effective problem solving which requires proper needs assessments to be conducted and strategies which address the relevant issues, to avoid unnecessary duplication of effort.

Numerous papers have been written worldwide which assess and advocate various water demand management interventions that play a significant role in decreasing water losses; however, few, if any, present a holistic perspective or take into account the importance of active community participation in the successful implementation of sustainable water demand management. Lack of community support has proven to be one of the key factors in the slow and inefficient delivery of services, often comprising the vandalism of much needed infrastructure and essential measurement tools, as well as a general disregard for the manner in which water is utilized. This leads to the failure of many potentially successful WDM projects.

5.1.2 Objectives of community awareness education

Given the importance of community engagement in any new WDM intervention, the objectives for such engagement must be clear. Communicating and awareness creation within the community is undertaken to achieve certain end results which should be clearly thought through at the start of the project. The following issues highlight why communication and education within the communities regarding water conservation are important.

5.1.2.1 Creating an informed public on efficient water use

It has often been said that knowledge is power, the premise being that education enables informed decision making and hopefully better choices regarding the subject of information. Understanding is a potent motivator of human behaviour and ensuring that people are properly informed regarding efficient water use can play a critical role in securing support from the communities. By communicating efficient water use, the old culture of bad practices comes into contact with the new efficient culture which catalyses the process of migration from one culture to the next. This of course is not an overnight process and changing attitudes may take years to achieve.

5.1.2.2 Establishing partnerships between the community and the municipality

One of the observed tragedies when it comes to the manner in which municipalities engage with communities, is the gaping hole of the 'us and them' dichotomy, which is often left un-bridged. Municipalities make large investments in installing smart devices to monitor and control water use and bill consumers, whilst communities invest time and energy in burning tires, placing large stones on the roads and creating imaginative placards; all aimed at cursing the municipalities for lack of service delivery. Although both parties have a unique and in many cases justified perspective on service delivery issues and what is required to effect change, more appropriate solutions which do not involve unnecessary expenditure or destruction of sorely needed infrastructure could be possible if a partnership is forged between the communities and municipalities. Communities are a valuable on the ground resource for municipalities which can be utilized to improve the state of the infrastructure through reporting of leakage for example, which is a form of passive leak detection. Education and awareness can thus help to establish such a partnership which must be strengthened through continuous engagement. Short-term and 'one-off' programmes will not be effective in the long run and such programmes will often create mistrust and scepticism if the communities feel that they are being exploited as a means to an end.

5.1.2.3 Promoting the municipality and its activities in the community

As mentioned previously, one of the key outcomes of the engagement between the municipalities and communities is not only to transmit information to people, which is the purpose of education and awareness, but also community participation which is geared at involving communities in decision making processes. WC/WDM by its very nature makes room for both forms of community engagement. Education is vital in helping people to understand the activities and projects the municipalities are implementing to reduce water losses but involvement in decision making is also vital so that the projects do not just remain the municipality's but become integrated into the lives of those they are meant to benefit. Classic cases depicting this kind of engagement involve two of the largest advanced pressure management installations in the world where schools competitions were organized to allow the community to select a suitable design for the chamber. In the case of the Khayelitsha installation, the original design shown in Figure 5.1 was replaced about 10 years later with the design shown in Figure 5.2.

Figure 5.1 Initial Khayelitsha installation shortly after completion.

By painting the chambers and drawing attention to them, they become a recognized land-mark to the communities they serve and by involving the communities in the design, they tend not to be vandalized. In the case of the large Sebokeng installation shown in Figure 5.3, the design was part of a schools competition where the pupil providing the winning design received a small prize, as did the school where the pupil was based (see Figure 5.4). At the time of writing this section, these two award winning installations had been running for 10 years and 13 years respectively, without any problems due to vandalism or theft and were, in fact, protected by the community members.

Figure 5.2 Khayelitsha installation after 'face-lift' (courtesy City of Cape Town).

Figure 5.3 The Sebokeng installation showing winning design.

Figure 5.4 Prize giving for the Sebokeng Installation design.

5.1.2.4 Promoting and enforcing water wise behaviour

The only manner in which the prevailing inefficient water use practices can be altered is through disseminating the correct information. Municipalities have a responsibility to educate consumers in respect of effective water use. The promotion of water efficient use must precede enforcement of municipal bylaws in order for consumers to understand the rules of engagement however; promotion alone is limited in effectiveness if not followed by enforcement. In many municipalities, the availability of human resources to enforce bylaws as well as the institutional knowledge to do so has posed a huge challenge and a serious impediment to the municipalities trying to comply with local regulations. This element is, however, crucial as without consequence, water conservation will remain in the realm of social responsibility as opposed to being seen as a responsibility and obligation every citizen must undertake.

5.1.2.5 Encouraging ownership of water loss reduction and infrastructure

The long term objective of WDM is to ensure that consumers take responsibility for water conservation and the infrastructure installed by the municipality. The water services are provided for the consumers, who are the direct beneficiaries of the infrastructure and as such; should play a role in reporting faults and directing municipal resources to where they are required. Communities are a valuable asset and possess local knowledge which can at times supplement or cover over limited institutional memory of municipal personnel. High staff turnover is experienced in many municipalities around the world which can pose difficulties in terms of locating and operating infrastructure for emergencies or routine maintenance. The participation of consumers is of great help in such cases to mitigate system failure through timely identification of problems.

5.1.2.6 Community awareness

As previously discussed, the implementation of community awareness is neither simple nor quick to achieve. The perception that public engagement, often labelled as the 'soft issues,' is a secondary consideration which need only be considered after other more important interventions have been completed, is not only a mistake but can be very dangerous. The recent dispute between Soweto and the City of Johannesburg concerning prepaid meters and their constitutionality or lack thereof, is a classic example of the problems that can arise due to a lack of proper communication and consultation with the community. In effect, a project with a budget of over $100 million was held up for several years due to the actions of angry citizens who felt that the proposed metering being offered by the City of Johannesburg was not in their best interests. With proper communication and education of the community many of the problem issues would have been resolved and the project would have proceeded as originally envisaged. Unfortunately, the community in Soweto mobilized their collective bargaining power and effectively stopped the project in its tracks for several years.

There are many other examples where costly investments have been made by municipalities with good intentions in order to increase cost recovery and water use control and in turn improve maintenance of infrastructure and service delivery. Many have been thwarted before completion and set the municipalities back years before they can recover from the debacles and resume efficient provision of services. These interventions are often well intentioned and well substantiated by the conditions of the municipalities, which can vary from old dilapidated infrastructure that must be refurbished, huge water losses with no means to recover the costs of service provision or mere shortage of resources, which necessitates restrictions. All these reasons for undertaking specific technical interventions are valid however, if the consumers do not know or understand these reasons, such interventions can merely come across as yet another ploy by the municipalities to make life difficult for consumers or to increase the cost of living. Proper consumer engagement is therefore critical in mitigating some of the difficulties previously discussed.

It is important to acknowledge that not all water loss or service delivery problems can be rectified through technical interventions. Figure 5.5 depicts a community problem in which crime has made a significant contribution to water losses in the area. The picture was taken in a neighbourhood where many brass taps had been stolen over one night, resulting in substantial water losses. This case was not the first of its kind in this community and surrounding areas. The immediate problem could be solved through replacing the taps, however, the root of the problem lies at the community level and it must therefore be addressed by the community. The community allowed a scrap-yard to exist which openly purchased stolen goods including brass taps. What was also interesting about this case was that the community members knew the perpetrators, and could pinpoint where the taps had been sold. Through engagement with the community policing forums in the area, the taps were located, recovered (see Figure 5.6) and eventually refitted, after which the theft of taps stopped. This example serves to highlight that proper community engagement can assist in rectifying the problems in partnership with the municipalities.

Figure 5.5 Impact of stolen taps on community.

Figure 5.6 Recovery of stolen taps by community support.

5.1.3 Methodologies for implementation

Each municipality is unique in some respect with a specific history, socioeconomic conditions and political landscape. Within this context there is no one size fits all intervention when it comes to community awareness and education. Lessons can be taken from other municipalities which have successfully implemented water loss reduction initiatives, however, the application of the methodologies will differ. Before implementation can commence, it is crucial for the implementer to ask a number of questions, which are as follows:

5.1.3.1 Why?

It is important to establish the purpose of the community engagement which must be undertaken. Is it geared at informing people regarding certain events, is it to educate the consumers or is it to involve them in a specific decision making process, to get their inputs or feedback? The 'why' will inform the 'how' of your intervention, in other words, what form of engagement will take place and what medium of communication will be utilized.

5.1.3.2 What?

The next step will be to establish the message to be disseminated to the consumers. If the goal of the engagement is specifically education and awareness, then it must be clear what the key messages will be. The idea is to focus on a few key issues and make that the overall theme of the education and awareness. The messages must also be based on proper research. The implementer must not simply assume the topic of engagement; the messages must be based on the reality of the community. There is no point telling consumers to save water in an area where consumers receive intermittent supply, for example. A more appropriate topic may be exploring other avenues of obtaining and storing water such as rain water harvesting.

5.1.3.3 Who?

The 'what' of the intervention discussed above will inform the 'who'. Once the pertinent messages have been established, the target audience for the message must be established which will in turn also determine how the message will be disseminated. It is crucial to be familiar with the demographics in an area such as prominent languages and ethnic cultures, literacy levels, levels of employment and which segment of the population is most acutely affected by the topic of education. For instance, there is no point simply handing out pamphlets with lots of writing on them if a large portion of the target audience is illiterate. Knowing the target audience will assist in tailoring existing material or creating new material to suit the situation.

5.1.3.4 How?

Based on the target audience, the implementer can now determine how the message will be disseminated. There are several methods which are commonly used for education and awareness purposes, which are as follows.

5.1.3.4.1 Door to door education and awareness

This method is particularly useful in cases where the levels of literacy in an area are low or a large segment of the populations has limitations in mobility, either due to age or availability and economic affordability of transportation. This method is also a fantastic vehicle for employment creation and community upliftment.

Identifying and utilizing people from the affected community to implement the programme is an excellent way of obtaining community cooperation, and assists in building trust between the community and the municipality. It also utilizes the existing local knowledge and skills. Bringing in personnel from another area may well be counterproductive and create more problems than it solves. This approach has been used successfully in many municipalities as depicted in Figure 5.7 and Figure 5.8. Door to door education and awareness also increases the visibility of the municipality in the community, which can raise consumer confidence and also present an easy channel through which to communicate issues to the municipality. This approach has many other added benefits including visual leak detection by those already on the ground moving from house to house and it is generally very cost effective as transport fees are usually not necessary if the programmes are ward based.

Figure 5.7 Community education and awareness.

Figure 5.8 Community education and awareness.

5.1.3.4.2 Public meetings and workshops

Public meetings, gatherings or organized workshops where whole communities are invited are also useful events for disseminating information. Such meetings are particularly useful in informing and educating large groups of people simultaneously, which may sometimes be necessary depending on the urgency of the information which must be disseminated. Usually, these gatherings are considered to be community events, which helps to create the necessary publicity and interest in the programme to trigger active participation from community members. This method of communication should, however, be used with care, depending on the sensitivity of the context. The topics of discussion must be made clear from the outset to avoid the meeting being taken over by the community to protest about other problems experienced by the community. Despite the possible problems, this method of communication is very useful and if conducted properly, will encourage communities to conserve water. Despite popular belief, communities are usually willing and able to participate in municipal activities and there is an observed increasing interest in engaging in water issues due to the importance of the resources, as depicted in Figure 5.9 and Figure 5.10, showing the attendance at recent water conservation workshops.

5.1.3.4.3 Pamphlets and posters and billboards

The use of pamphlets, posters and billboards is common in education and awareness campaigns and very effective if utilized appropriately. In the development of such material, care must be taken that the information is presented in a manner which is comprehensible to the target audience and free from sensitive issues such as race and religion. This includes using the appropriate languages and making sure that the writing is simple, clear, uncluttered, precise and minimal. Using pictorial illustrations is often very helpful in communicating the message and helping people to relate to the information presented.

Figure 5.9 Public meetings.

Figure 5.10 WDM workshop.

5.1.3.5 When?

Municipalities wishing to engage with the community must be realistic regarding the time frames in which they expect to see results. Building relationships and trust with communities often requires commitment and persistence. It is important to note that community awareness interventions are not short term 'hit and run' initiatives but rather long-term strategies to address water losses. Wasteful habits that have been entrenched over years and decades will require months or years to alter.

Implementers should also be mindful of what is happening in the municipality before commencing with such programmes. It is not practical, for example, to initiate a community awareness programme shortly before national or local elections, as this is often seen as a ploy to capture votes for certain political parties. It is thus crucial that such interventions are appropriately timed and given sufficient time to properly monitor the process.

5.1.3.6 Schools awareness and leak reduction

Schools are an essential part of any WDM education and awareness programme. Instilling good values and habits at the foundation developmental stage is invaluable as the future leaders will start practising wise water use behaviour and encourage others to act in a similar manner. Young learners are particularly receptive to WDM education and awareness as they have often not yet developed the inefficient habits practiced by adults. Based on schools programmes undertaken previously, some of the key challenges and issues facing schools with regards to WC/WDM are as follows:

5.1.3.6.1 High levels of leakage

Water loss through leakage appears to be a serious problem for schools, particularly those found in underfunded and developing areas such as shown in Figure 5.11, Figure 5.12, Figure 5.13, and Figure 5.14. A number of the schools visited, exhibited high levels of visible leakage which can usually be quickly addressed. Maintenance of internal plumbing fittings appears to be a world-wide challenge due to ongoing vandalism and theft of water fittings. This is often aggravated by insufficient funding, and poor workmanship by the plumbers.

Figure 5.11 Leaking toilet in school.

Figure 5.12 Vandalized toilet in school.

Figure 5.13 Leaking tap in school.

Figure 5.14 Leaking toilet fitting in school.

The most significant source of water losses in most schools is through the toilets, some of which can leak profusely for months on end if not years. On the positive side, however, schools often express their desire for greater support and guidance from both the municipalities and the District Department of Education to address these challenges. It is however important to note that in many municipalities, a large portion of the schools are allocated with a maintenance budget which can be utilized at their own discretion. Many schools complain that the budget is not sufficient to address all the maintenance needs, however, the majority of leaks found in the schools do not require huge capital expenditure. In most cases, simply replacing the tap and toilet washers suffices.

Automatic flushing urinals also waste huge volumes of water and have already been outlawed in many parts of the world. Despite being illegal, such toilets remain in operation in many areas and continue to waste water, as shown in Figure 5.15 and Figure 5.16. The presence of these urinals usually results in excessively high water bills even during school holidays as they continue to flush regardless of actual use.

Figure 5.15 Automatic flushing urinal in school.

Figure 5.16 leaking urinal in school.

5.1.3.7 Inaccurate billing and broken meters

Billing and metering is a contentious issue in many municipalities and if not conducted properly, negatively affects the willingness of consumers to pay for services. Before the commencement of the assessments undertaken in schools in previous programmes, one years' consumption data should be requested from the municipalities for all the schools participating in the programmes. In many cases it is found that the billing is based on estimates of water use rather than recorded meter readings. In such cases, meter investigations should be undertaken to assess the extent of the problem and all faulty meters should be replaced immediately. Typical meters in very poor condition are shown in Figure 5.17 and Figure 5.18. In order to improve the communication between the schools and the municipalities, and to ensure effective cost recovery, municipal response to reported problems must improve.

Figure 5.17 Leaking school meter.

Figure 5.18 Illegible school meter.

5.1.3.8 Poor response and communication with the municipality

As mentioned previously, communication between the municipalities and the schools can be a problem and is often lacking. The lack of responsiveness or a delayed response to billing related queries also contributes to tensions existing between the schools and municipalities. A coherent support structure is required from the municipalities to assist the schools in resolving billing matters.

5.1.3.9 Relatively low levels of water conservation awareness

Based on the results from the score card assessments undertaken in numerous schools, it is evident that more intensive water conservation and awareness education is required to reduce the water losses and to change the inefficient consumption patterns in some of the schools. General observations made during the course of the assessments included the following:

- Irrigation of school gardens during the midday heat using potable water
- Leaks left unattended for long periods of time
- Very limited rain water harvesting
- School management with no knowledge of the location of the school meter
- Frequent occurrence of internal plumbing leakage due to vandalism
- General wastage of water by learners (see Figure 5.19)
- Poor hygiene conditions (see Figure 5.20)

Figure 5.19 Inefficient water use by learners.

Figure 5.20 Unhygienic conditions in schools.

In general, it has been observed that water conservation is a culture which must be cultivated in the schools. The state of the schools often reflects the leadership of the schools and the participation of the whole school community including principals, educators, learners and school governing bodies and parents is required. Whilst some schools display inefficient water use and poor hygiene conditions as seen above, others within the same socioeconomic conditions exhibit exemplary practices a seen in Figure 5.21 and Figure 5.22, and it comes down to the efforts and leadership of the parents, educators and school management where the same message is communicated to learners at school and at home.

Despite the seemingly dire picture painted by the challenges outlined above, it must be noted that there are many schools which are implementing admirable water conservation strategies including: keeping buckets of water in the classrooms to increase monitoring during teaching time, use of boreholes for irrigation of food gardens used for feeding schemes, rainwater harvesting and self-initiated water conservation workshops on spring days when children would habitually play with water to celebrate the day. These have been some of the outcomes of the schools WC/WDM campaigns implemented previously and illustrate the value of engagement with schools. It is recommended, however, that such programmes should not focus only on the learners. Water conservation at school is a team effort and can only be sustained through the participation of all involved. The participation of the Department of Education is also invaluable. In some countries it is found that the local Department of Education has established a system where schools are obligated to provide their own meter readings on a monthly basis. This is an excellent approach to raising the profile of water management in this sector and helps to raise the awareness of decision makers in the schools, who start to pay attention to the consumption water use in the schools.

Figure 5.21 Water wise mural at school.

Figure 5.22 Efficient drip irrigation.

In undertaking schools awareness campaigns, the following should be considered:

5.1.3.9.1 Whole school participation

Each segment of the school community has a crucial role to play in water conservation, and as such should be included in the process. As mentioned previously, the focus should not only focus on the learners but also include the principals and governing bodies as the school management structures which set the rules and can help to drive the WDM initiatives. The general assistants who deal with maintenance issues on a daily basis should be included in all campaigns. One of the most common complaints within schools is the lack of urgency in fixing leaks, which tend to run for months if not years. Most of the leakage does not require highly technical expertise and can be addressed by the general assistants if some very basic training is provided. The provision of such training is very useful and should be included in any schools programme. Similarly, the educators should be included since they pass on their knowledge to the learners. It is thus crucial that the educators are provided with the knowledge and tools to promote WDM to the learners.

5.1.3.9.2 Schools competitions

Schools competitions have been found to be very effective motivators of water conservation in schools. They allow both learners and educators to engage vigorously with the water conservation issues and to compare their performance against that of other schools, which in itself creates greater awareness.

5.1.3.9.3 Use of media in creating awareness

The media is powerful tool for driving public participation as well as education and awareness. It engages a variety of senses and if utilized well, can allow the public to interact with projects and with municipalities and provide feedback on their concerns and issues. The numerous forms of engagement mentioned previously primarily target individuals who are at home during the day and thus greatly impact on the peak demand periods. In water conservation and demand management related campaigns, however, it is just as critical that household owners who are paying for water services also be reached. This segment of the population is typically at work during the day and forms the very important tax base which contributes to the economic viability of the country. Door to door education and awareness or workshops held during the day cannot reach this target market therefore other mediums of communication must be considered. In this respect, local community radio stations can be used to target the working population. This type of media is widely accessible to most people, and covers all bases in terms of literacy levels, because no reading is required. Local newspapers can also be used to reach the working population and are a popular and affordable means of communication both for the target consumers and implementing agents.

One of the most potent and effective forms of communication is visual. When educating consumers on water conservation, it has been observed that people respond very positively to mediums such as power point presentations where pictures are shown of water wastage, which must be altered. Television is thus also a very effective medium for disseminating information as it combines both visual and auditory mediums of communication. In South Africa, the Government Department of Water Affairs actively promoted a 'War on Leaks' campaign on national television. The advertisement had to be removed shortly after it was aired due to the massive response from the public in which callers from all over the country reported leaks to the Department. As the repair of such leaks is the responsibility of the municipalities, the Department of Water Affairs found itself in a difficult situation and the 'report a leak' campaign was hastily cancelled. This example did highlight the potential value of using television to help address leakage and water wastage.

Irrespective of the mediums used to communicate, the following should be taken into consideration to ensure that the media is used effectively:

5.1.3.9.4 The message should be clear and uncluttered

Due to the fact the media is such a powerful tool for communication, it also has the potential to create confusion and the message can be distorted if not formulated properly. It is thus crucial to the keep the messages short, concise, clear and to the point.

5.1.3.9.5 Simple language

The educational material, particularly for written media, must be easily understood. The idea is to keep the language simple and un-convoluted to ensure that the messages are received loud and clear.

5.1.3.9.6 Focus

WC/WDM is a very wide field which involves a multitude of topics ranging from rainwater harvesting to pressure management. In education and awareness it is crucial to focus the themes or topics for the awareness, based on the crucial issues identified in the community in order to aid assimilation of the information. Consumers must be left with a few key messages to process at any one time. The education and awareness can therefore be structured and undertaken in phases in which the messages can be changed periodically.

5.1.3.10 Role of politicians and councillors

Political support of any water conservation and demand management projects is essential for their success. Many a project has seen success or failure on the basis of the participation of councillors and ward committee members and their support or opposition will generally make or break a project. Ward Councillors or elected cultural representatives should be the starting point in gaining entry to any community. Communities also look up to these individuals as the example and expression of democracy and their right to choose and participate in the affairs of the land. It is thus crucial that these individuals are engaged and involved at the commencement of any community based programme to avoid costly and unnecessary project delays.

This issue is particularly challenging for most municipalities since it often comes down to a choice between the popular decision and the correct decision. Many politicians will rather take the popular decision since if they opt for the unpopular choice, they may well not have a job to come back to in 12 months time. The fact is that such difficult decisions can be made and successfully implemented with proper consultation and strong leadership from the politicians and leaders.

5.1.3.11 Monitoring and evaluation

One of the main concerns related to education and awareness has frequently been how these initiatives can be monitored and evaluated. It is important to remember that communication is not a one way stream of information. It depends on a cycle of input and feedback. In this regard, undertaking Knowledge Attitudes and Perception (KAP) Surveys can be a useful tool. Although not without their flaws, if they are undertaken prior to the commencement of a community awareness programme at the mid-period of implementation and at the conclusion of the programme; they can provide very valuable feedback regarding the effectiveness of the programme and how the communities feel about the initiative.

In cases where an area is discrete and fitted with a bulk meter, it becomes technically possible to monitor the impact of the interventions through the bulk meters, provided there are no other interventions being undertaken at the same time. Figure 5.23 shows the consumption patterns in such an area, where a community awareness programme was implemented in the absence of other technical interventions. Through community awareness conducted over a 12 month period, there was a reduction of 6 m^3/h in the minimum night flow as residents began to fix their internal plumbing leakage.

The main benefit of community awareness programmes is to enhance and assist with the sustainability of technical interventions. Their true value lies in opening the channels of communication between communities and municipalities, building a culture of ownership of municipal projects and thus protecting costly investments in infrastructure. It is indeed very difficult to quantify the potential savings which can be made through such interventions, however, the benefits are often acknowledged retrospectively when damage control is conducted on projects which have been terminated due to lack of engagement. The City of Cape Town is one major city where some very detailed and comprehensive education and awareness interventions were introduced together with a detailed auditing system which was used to measure the effectiveness of the programme. It is one of the few case studies in the world where such a detailed and rigorous system for monitoring and evaluation has been implemented. The results were very encouraging and have been presented at a number of international conferences.

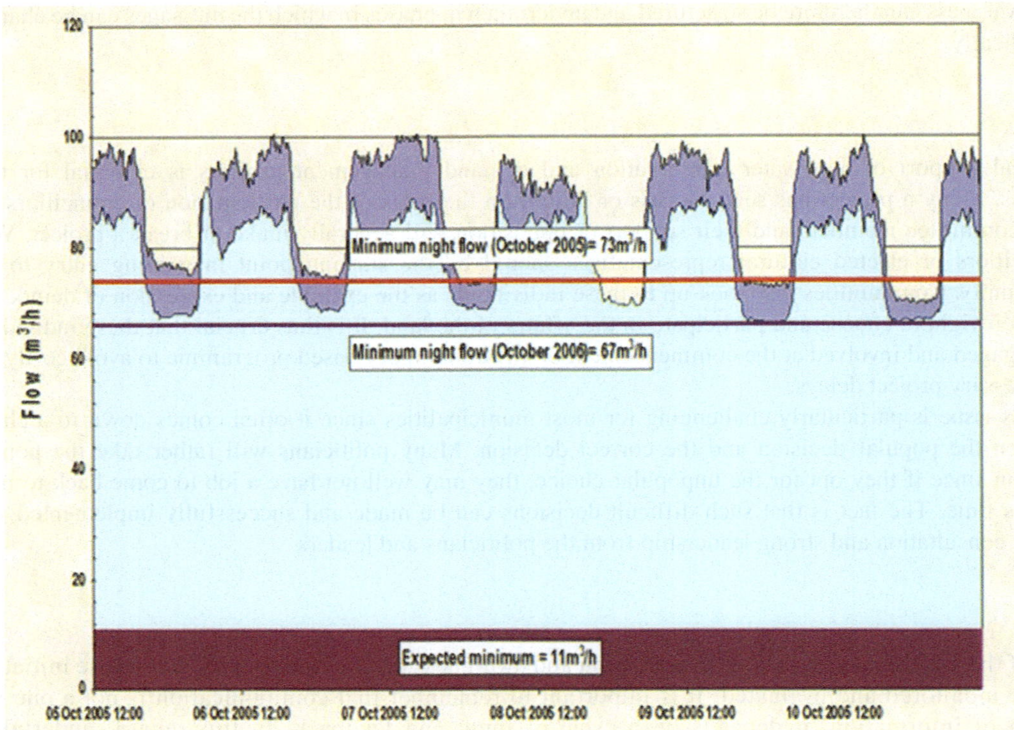

Figure 5.23 Graph of consumption in a discrete zone.

5.1.4 Lessons learnt

5.1.4.1 Retain capacity

A key challenge in many of the municipalities is how to retain the capacity developed during the implementation of water projects. The projects are often implemented as short-term solutions with little thought for their long-term sustainability. Should the status quo remain unchanged, little progress will be made in improving the delivery of services and the same problems will continue to persist. Money spent on perpetual training programmes can be more effectively utilized to retain the individuals already trained and directed to the implementation and operation of other water demand management measures that are required to minimize the water losses.

5.1.4.2 Clear vision and leadership – relationship between the service provider and the municipality

Retaining capacity without clear vision and leadership can prove fruitless. The resources developed through water conservation education and training must be utilized for their intended purpose and for this to take place, proper guidance and continuous monitoring and evaluation from the municipality leadership is required. If this can be achieved, there is no reason why lack of capacity should remain an obstacle to service delivery.

5.1.4.3 Consistency and sustained effort

Both training and capacity building and community awareness programmes share one common condition for success; they both require consistent and sustained effort. So often, momentum is built up during the implementation period of the projects, only to be deflated in a matter of a few months when the implementing agents leave the system. This is a concern because unlike technical programmes, initiatives that involve people cannot be easily rectified or recommissioned. An extensive investment of time is required to open the channels of communication between communities and municipalities, but more importantly they also require a degree of trust and goodwill. The culture of short term hit and run interventions undermines open communication and the process of establishing efficient water use practices. If these interventions are undertaken in such a manner, the communities cannot be expected to display any long-term changes in behaviour. Words must be supported by actions, with the municipalities setting the example. Likewise, building capacity requires mentorship and consistent effort to establish the experience necessary for continued successful implementation of water demand management initiatives.

5.1.4.4 Communities are an essential resource

Demand management is often conceptualized as a model comprising a wide range of technical interventions requiring sophisticated methods and equipment. Communities are an essential resource for the municipality, notwithstanding the fact that 80% of the implementation of demand management takes place at grass roots, in individual homes where the consumers must adapt their water use practices to promote economic efficiency and sustainability of the resource. They must be taken on board and provided with the necessary skills to take ownership for their progress.

5.1.4.5 Clear maintained communication

The provision of clear realistic information is essential to healthy communication between municipalities and communities. Particularly when it comes to the provision of basic services such as water, which is often a fairly sensitive issue, unrealistic promises of free services and short delivery times are neither helpful nor sustainable. Effective Demand Management requires the provision of consistent, realistic and honest information to consumers that allows them to make informed decisions.

BIBLIOGRAPHY

Siqalaba Z., Kekana C. and McKenzie R. (2006). Reducing water wastage in Sebokeng and Evaton, *IMESA Conference*, 11–13 October, Soweto.

Siqalaba Z. and Wegelin W. (2008). Emfuleni Water Loss Project – Public Participation Programme, SAICE Awards, Midrand, Ghallager Estate, Johannesburg, South Africa.

Siqalaba Z. and Wegelin W. (2009). Public Participation in Water Demand Management. Paper presented at the *IWA Water Losses Specialist Group Bi-Annual Conference*, City of Cape Town, 26–30 April.

Siqalaba Z., Wegelin W. and Mckenzie R. (2010a). Community Management Proves Key to Successful Demand Management. International Water Association, *Water 21*, August, pp. 61–62.

Siqalaba Z., Mckenzie R. and Wegelin W. (2010b). The Role of Training and Capacity Building in the Implementation of Water Demand Management. *Water Institute of South Africa, Bi-annual Conference*, International Conference Centre, Durban, 18–22 April.

Chapter 6

Meter logging and recording

6.1 INTRODUCTION

6.1.1 General

With the rapid developments in internet based technology, the value of using the internet as a platform for data access and display is being recognized throughout the world. Remote logging through GSM and GPRS devices in particular are commonplace and easily accessible to all water suppliers in virtually every country. The costs associated with both the hardware as well as the software and associated communication charges have all decreased in recent years with the result that the financial viability of using such technology to assist with leakage reduction activities is now clearly evident (Mckenzie, 2014).

This chapter will provide a brief overview of the use of flow and pressure loggers in a water reticulation system to highlight the value of such information as part of any leakage reduction programme. It will also present details and interesting examples from a wide variety of case studies to highlight the key issues associated with both flow and pressure logging.

6.1.2 Importance of logging

Logging involves recording pressures and/or flows entering a zone and can be compared to a doctor recording the blood pressure and heart rate in order to determine the health of a patient. From a management viewpoint, the logging results often highlight problems within the water supply system which can then be addressed. Without regular logging results and associated analyses, such problems can exist unnoticed for months if not years and result in higher leakage and/or poor service delivery. It is the view of the authors, that regular logging of both flows and pressures are an essential component of any well managed water supply system.

Having stressed the importance of undertaking regular flow and pressure loggings in a water supply system, we note that it is equally important for the technician or manager receiving the information to understand what it represents. Analysing and interpreting logging results is often very confusing and difficult and can be compared to 'detective work'. The remainder of this section is devoted to the interpretation and understanding of logging results and all graphs shown are from real examples collected by the authors.

Logging of pressures and flows is one of the most important aspects of any WDM programme and one that is generally neglected if not ignored completely. In recent years, with the advent of GSM and GPRS based loggers it is now possible to capture and transmit logging information with relative ease. Such information, if used properly, can help water managers to manage their water distribution systems effectively and efficiently and they can often identify problems before they become major crises.

There is no substitute for reliable 'real-time' flow and pressure information and where available, it will facilitate the analysis of Minimum Night Flows which was the original foundation of any proper WDM programme. Over the years, more effort seems to have been placed on high-tech electronic solutions and software which eliminates the need to visit the site or to get one's hands dirty. Recording and analysing Minimum Night Flows, has in fact become less common as the new technology and software models eliminate the need for any such analyses. In reality, many water reticulation managers have lost the ability to properly manage their systems in many parts of South Africa and there is a real need to move back to the basics and start using logging results to monitor the water reticulation systems and to find problems. Reliable logging information provides a manager with real-time data on what is really happening in the system as opposed to what the hydraulic models say should be happening. In most cases, the modelled flows and pressures bear little resemblance to what is being measured on the ground. Such discrepancies are not a reflection on the software models that are used to design the reticulation systems. Such models are only as reliable as the information on which they are based. When the models and reality diverge, the problem is a reflection on the reticulation managers and maintenance personnel who are not operating the system in the manner in which it was designed.

In the remainder of this section, the value of real-time flow and pressure logging information will be discussed and many actual flow and pressure logging results will be examined to highlight some of the common issues that will be found in most water reticulation systems.

6.1.3 Interpretation of minimum night flows

The minimum night flow entering a zone is one of the most important and useful indicators for assessing potential leakage in an area. For it to be useful, the zone in question must be properly sectorized and discrete. In very general terms, the minimum night flow, is the flow entering a zone during the time at which most of the normal legitimate water use is at its lowest. During the period of minimum night flow, it is often (not always) possible to estimate the likely level of legitimate night use as well as the background leakage which cannot be eliminated. Both components are to be expected and can be added together to provide an estimate of the expected night use. Once this expected night use has been estimated, it can be compared to the actual recorded night flow. If the two are reasonable similar, then it suggests that there is little unexplained leakage in the zone and if the recorded night use is significantly higher than the expected value it suggests that there is a significant amount of unexplained leakage that should be investigated. This is effectively the basis for the well-known Burst and Background Estimate (BABE) methodology as developed by the UK water industry to identify zones with potentially high leakage levels. As the name suggests, it is based on certain estimates which are often unknown and therefore may be overestimated or underestimated. In general, however, it is usually found that the methodology is relatively robust and even if one or two of the estimated parameters are inaccurate, the overall result remains useful in identifying areas with high leakage.

6.1.4 Assessing the minimum night flow

Many simple models have been developed throughout the world using the BABE methodology to analyse minimum night flows and are usually freely available through the internet. One such model is the

SANFLOW model developed through the South African Water Research Commission, which is fully discussed in the original *SANFLOW User Guide* (WRC, 1999). The user guide provides a comprehensive explanation of the BABE methodology for analysing the minimum night flow in an area, and a brief summary of the approach is provided below for reference.

The measurement of background night flows is one of the most important actions that can be taken to identify leakage problems. It is often possible to identify many problem issues by simply looking at the minimum night flow. The minimum night flow is usually found to occur sometime between midnight and 4 am when the consumption in the network is at its lowest. In some countries, the usual approach is not always applicable especially in areas where many residents or municipalities irrigate at night with automatic sprinkler systems or in areas where there is continuous water-use activity throughout the night. The approach developed un the UK is suited to normal residential or industrial areas that have little active water use between midnight and 4 am – the flow entering such an area is shown in Figure 6.1.

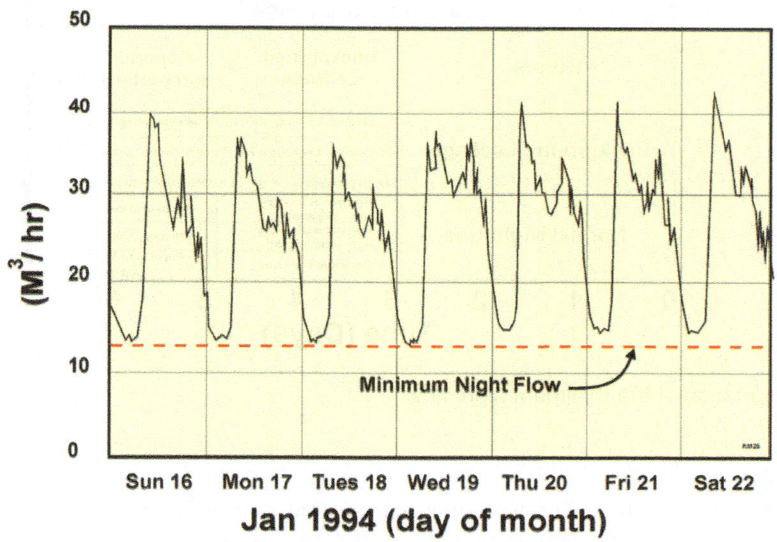

Figure 6.1 Typical flow logging for zone showing minimum night flow.

Having logged a zone meter in order to establish the minimum night flow, it is then necessary to establish the various components of the night use in order to estimate the level of unexplained leakage. The key issue is to establish whether or not the zone has a problem or if it is healthy and has no problems. Through the use of the BABE methodology, the minimum night flow is considered to consist of three main components namely:

- Normal legitimate night use
- Background losses
- Burst pipes.

This breakdown is shown in Figure 6.2 from which it can be seen that the normal use and background losses have each been further divided into three smaller components which can be estimated using certain simple and basic assumptions that tend to be surprisingly close to what is found in the field. The normal night use and background leakage can be estimated for any specific zone given a few basic parameters which are either measured or estimated depending on the level of information available for the specific zone. After logging the flow into a zone for a few days or weeks to establish a realistic estimate of the minimun night flow, the unexplained leakage, which is due mainly to bursts, can be determined by simply subtracting the estimated background leakage and normal night use from the measured minimum night flow.

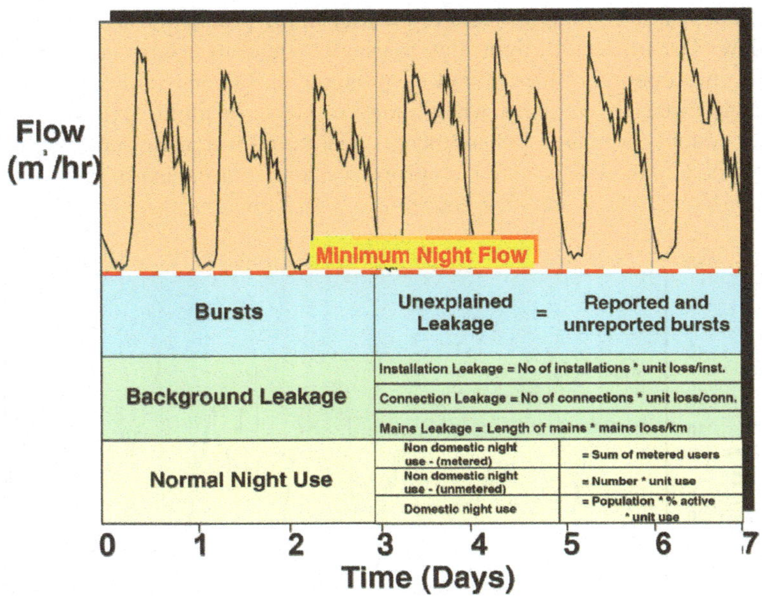

Figure 6.2 Interpretation of the minimum night flow.

6.1.5 Normal night use

The expected legitimate water use in a zone cannot be measured accurately every evening since it comprises a large number of small users – sometimes more than 2000. As it is impractical to try and measure the use for each small user during the period of the minimum night flow analysis, it is estimated, based on certain norms and variables that have been refined over the years through practical research around the world.

For the purpose of the analysis, the expected normal night use is split into three components namely:

- Normal domestic night use – mainly toilet flushing
- Small non-domestic night use such as hotels and garages, and so on.
- Larger users such as hospitals or factories and so on (recorded individually).

6.1.5.1 Normal domestic night use

Normal domestic night use represents the water used during the night in a household and is predominantly due to toilet use. Use of water for making coffee or tea represents a very small portion of the overall household use and is effectively ignored. Experience in various parts of the world has shown that approximately 6% of the population are active during each hour and that the water used per toilet flush is in the order of 10 l. Based on these estimates the normal domestic night use is estimated.

Normal night use = 6% of population $*$ 10 litres per hour

This may seem over simplistic, which is true, however, it usually provides a reasonable estimate of what to expect in an area and will most likely err slightly on the high side. It should be noted that in areas with high internal plumbing losses, the water being 'used' at night in the households will be much higher than the estimated normal night use. If this is the case, the analysis of the minimum night flow will suggest very high burst leakage – much of which will in fact be internal household leakage. The purpose of the analysis is to identify if something is wrong and if the assessment shows that there is very high burst leakage, then it must first be checked to establish if, in fact, the leakage is burst leakage or a combination of burst leakage and household leakage. Checking the sewer flows at night will often answer this question.

6.1.5.2 Small non-domestic night-use

The small non-domestic night use is more difficult to evaluate and depends to a large extent on the type of businesses (if any) operating in the zone. The assessment is not ideally suited for industrial areas or residential areas which have a high component of industrial use at night. Although each small non-domestic user is metered individually it is again impractical to record each of the meters during the night flow exercise. Instead, the users are lumped into various categories and a typical night use is assumed for the group. For example, there may be several all-night garages or all-night cafes where the unit use is relatively small, although when added together the total use may be significant. A range of typical night use values for various different commercial enterprises has been produced based on extensive studies undertaken overseas and guidelines are provided in the *SANFLOW User Guide.*

6.1.5.3 Large non-domestic users

In some zone metered areas, it is often found that there may be one or more large water consumers whose consumption can have a significant influence on the night flow analysis. In such cases it is necessary to meter such consumers individually to determine how much water has been used during the night flow exercise. Only one or two such consumers would reside within a specific zone and therefore they can easily be checked during the night to assess exactly how much water they consume during the period of minimum night flow. Consumers falling into this category would include airports, large hotels, breweries, swimming pools, and so on.

6.1.6 Background leakage

Background leakage is the cumulative leakage from all relatively small leaks in the reticulation system. Such leaks occur from valves, joints, hydrants, stop-taps, meters, dripping taps, toilet cisterns, roof tanks, and so on. Individually such leaks are generally uneconomic to find and repair with the result that background leakage is accepted as a fact of life in all water reticulation systems within certain limits.

In general, background leakage can be split into three main components namely:

- Background leakage from each kilometre of mains
- Background leakage from each connection
- Background leakage from each property.

6.1.6.1 Background leakage from mains

There will always be some background leakage from any distribution system, some of which occurs from the water mains. Small leaks often occur at the pipe joints or from small cracks or holes in the pipes and the magnitude of the leakage is dependent upon the condition of the infrastructure and of course the operating pressure. For the purpose of the assessment of the minimum night flow, all leakage parameters used in the calculation are based on a standard operating pressure of 50 m. When assessing the expected level of background leakage from systems operating above or below this pressure, the calculated background leakage must simply be adjusted upwards if the pressure is greater than 50 m and adjusted downwards if it is operating below 50 m. The adjustment for pressure is fully discussed in the *SANFLOW User Guide* and is not repeated in this section. Typical background leakage rates for normal diameter water mains suggest average values of around 40 l/km of mains per hour. A new pipeline may have lower leakage, sometimes even below 20 l/km/hr while older pipelines can have much higher leakage rates of 100 l/km/hr or even higher.

6.1.6.2 Background leakage from connections

Poor workmanship coupled with general wear and tear often results in leaks from pipe connections. In general, we usually work on the basis of one pipe connection to each property, although in many systems a single connection can sometimes be used to supply multiple properties. This issue has been discussed and debated at many venues and in the view of the author, it is appropriate to assume that the number of connections or 'equivalent connections' is equal to the number of properties except in cases of apartments where the actual number of connections should be used. Connection leakage is considered as the leakage occurring from the connection at the water main to the water meter at the property or to the property boundary in cases where no meters exist. In most water distribution systems, the connection losses are often the major source of loss from the system. For most purposes, a value of 3 l per property per hour can be used which will provide a reasonable estimate of the expected background property leakage.

6.1.6.3 Background leakage from properties

The third and final component of background leakage used in the calculation reflects the property leakage for each property after the consumer meter – as opposed to the previous component which relates to leakage in the connection pipe before the customer meter. This component is intended to allow for some leakage in the pipe from the meter to the property and also some leakage inside the property. Under normal circumstances, an allowance of 1 l/property per hours is appropriate.

A common criticism of this component based methodology is often raised concerning areas where the most significant source of leakage is from the plumbing fixtures inside the properties. This has been mentioned previously but as it is such an important issue it deserves further explanation.

In South Africa, household leakage is often the most serious leakage issue in the zone and the household leakage can dominate the whole water balance and therefore also the Minimum Nigh Flow. In such cases, the household leakage will not be close to 1 l/property per hour but may be several hundred litres per property per hour. In such cases, all of the other components of the Minimum Nigh Flow analysis are irrelevant and the analysis suggests that there is a huge burst problem. On closer examination, it will be found that there is no sign of such high burst leakage and from experience, the individual looking at the analysis results will often be able to confirm that the problem is most likely inside the properties. In reality, this is not a difficult issue to identify and in cases where household leakage is found to be the problem, it will usually be so significant and clear cut that there is no question about where the problem lies.

6.1.7 Calculation of unexplained bursts

Having measured or estimated the various components of normal night use and background night use, the two figures are added together and then subtracted from the measured minimum night flow. The difference is the unexplained losses that are attributable to either unreported bursts or to errors in the assumptions made during the calculation.

To demonstrate the analysis of a minimum night flow using the BABE methodology, it is easier to make use of a simple example. In this example, a case will be used in which the average zone night pressure is at 50 m, which is the base pressure at which no pressure correction factors are required. The basic information required for the calculation is provided in Table 6.1.

Table 6.1 Basic information needed for MNF analysis.

Description	Value
Length of mains	9300 m
Number of connections	600
Number of properties	672
Estimated population	3000
Average zone night pressure (AZNP)	50 m
Measured minimum night flow (MNF)	14.4 m^3/h
Background losses from mains	40 l/km/h
Background losses from connections	3 l/connection/h
Background losses from properties	1 l/connection /h
% of population active during night flow exercise	6%
Quantity of water used in toilet cistern	10 l
Number of small non-domestic users	30
Average use for small non-domestic users	50 l/h
Use by large non-domestic users	1.2 m^3/h

Table 6.1 also includes the various default parameters that are normally used to calculate the background leakage and the normal night use as mentioned previously. Having established the default loss parameters, it is now possible to estimate both the normal night use and the background leakage. The respective calculations are provided in Table 6.2 and Table 6.3.

Table 6.2 Estimate of normal night use.

Description	Calculation	Value
Domestic night use	3000 @ 6%/h @ 10 l	1.8 m^3/h
Small non-domestic use	30 @ 50 l/h	1.5 m^3/h
Large non-domestic use	1 @ 1.2 m^3/h	1.2 m^3/h
Total normal night use		4.5 m^3/h

It should be noted that a pressure correction factor is indicated in the above table. In the case of this example the operating pressure is known to be 50 m which is considered to be the standard pressure. At standard pressure, no pressure corrections are required and it can be seen that the pressure correction factor is calculated to be 1.0 (i.e., no change). The topic of pressure correction is discussed in the *SANFLOW User Guide* (WRC, 1999).

Table 6.3 Estimate of background leakage.

Description	Calculation	Value
Mains losses	9.3 km @ 40 l/km/h	0.37 m³/h
Connection losses	600 @ 3 l/connection/h	1.80 m³/h
Property losses	672 @ 1 l/property/h	0.67 m³/h
Total background leakage at 50 m pressure		2.84 m³/h
Pressure correction factor	$(50/50)^{1.5}$	1.00
Total background leakage at 50 m pressure		2.84 m³/h

Having estimated the two main night-time water use components, it is possible to calculate the difference between the measured minimum night flow and the expected legitimate night time use. Table 6.4 provides the calculation that identifies the level of unexplained leakage in the given zone metered area.

Table 6.4 Estimate of burst leakage.

Description	Value
Expected background leakage	2.84 m³/h
Expected normal night use	4.50 m³/h
Total expected night use	7.34 m³/h
Measured minimum night flow	14.40 m³/h
Unaccounted-for leakage (14.40–7.34)	7.06 m³/h

As can be seen from the table, it is estimated that in this example the unexplained leakage is in the order of 7 m³/h. The remainder of this section is devoted to the interpretation of various real logging results to demonstrate the value of such information and to assist those new to such logging results to understand and interpret the various examples, each of which highlights a specific problem.

6.1.8 Using minimum night flow for leakage management

This is possibly one of the most important sections in the book and it will highlight the value and potential for using the logged minimum night flow in an area to assess the likely physical leakage as well as the potential for improvement.

The first figure to be shown is that of the flow measured entering a zone over a 24-hour period, as shown in Figure 6.3. The flow logging result for the same zone over a 3-day period is shown in Figure 6.4. As can be seen from Figure 6.4, the 3-day logging result indicates a typical domestic demand pattern which is repeated each day and has a morning and evening peak demand together with a minimum demand sometime during the late evening or normally around 02h00 in the morning. Although it is possible to derive some information from the 24-hour graph in Figure 6.3, the 3-day logging result is of greater value and where possible a 10-day logging result is preferred.

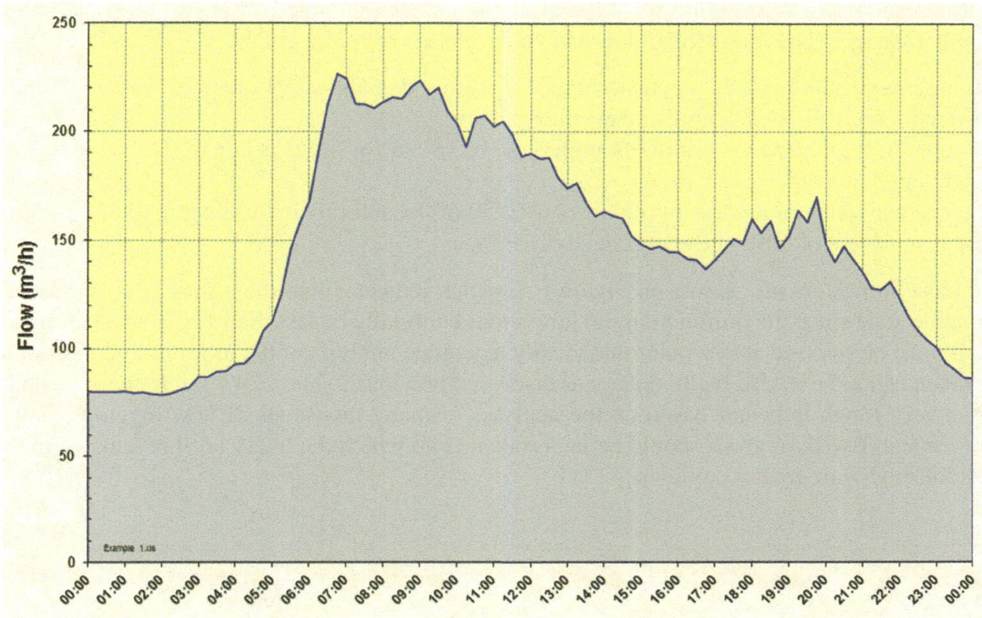

Figure 6.3 Flow entering a zone over a 24-hour period.

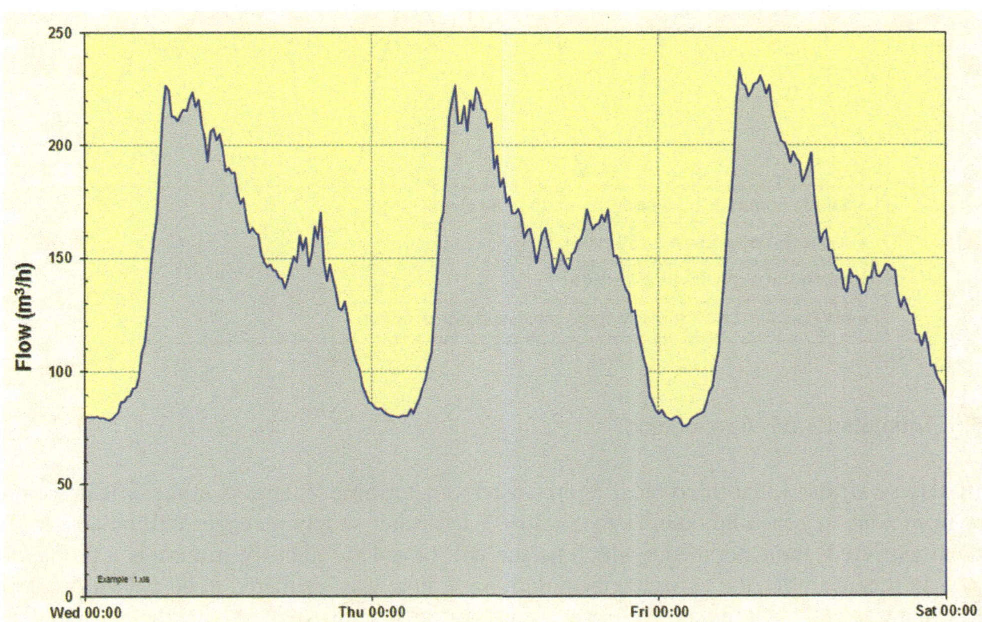

Figure 6.4 Flow entering a zone over a 3-day period.

From the results shown in Figure 6.4, several key points can be noted which are of great help to the water supply manager. The following points can be concluded from the graph:

- the logging results indicate a consistent and repeating demand/supply pattern which is typical of a residential area with a morning and evening peak.
- the logging results show a minimum night flow of ±80 m^3/hr – this is one of the key indicators that water supply managers wish to see
- The average daily demand is in the order of ±150 m^3/hr and this indicates that the minimum night flow is in excess of 50% of the average daily flow.

These basic indicators are shown on Figure 6.5, which indicates that there may be a serious leakage problem in the zone since the minimum night flow should normally be less than ±20% of the average daily demand. It must be stressed at this point that ideally a detailed analysis of the minimum night flow should be undertaken using the BABE methodology as discussed previously. The ±20% guideline provided above is simply a very rough indicator based on the analyses of many thousands of flow logging results. If in doubt, a thorough BABE analysis should be undertaken from which the likely level of leakage in the area can be established with greater confidence.

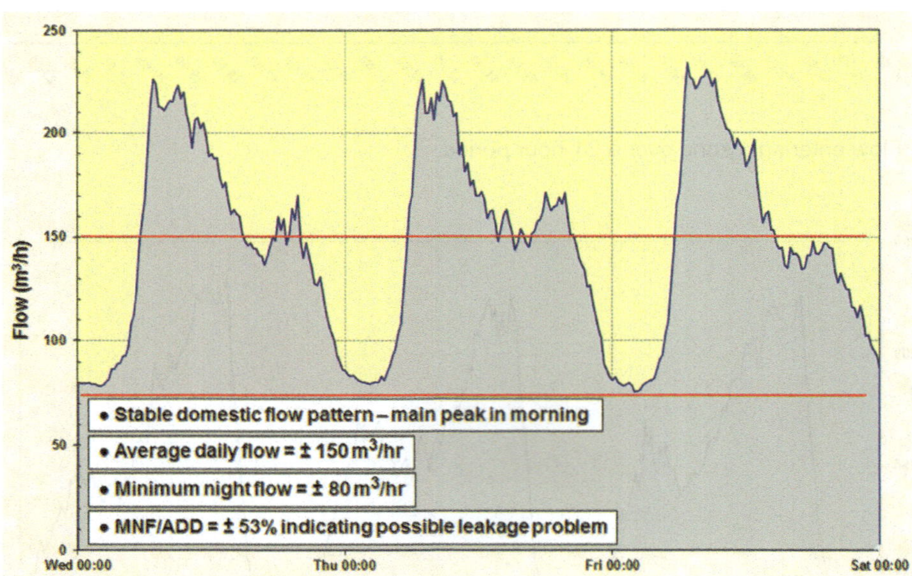

Figure 6.5 Annotated 3-day flow logging.

It must also be stressed that the typical figures used to determine if there is in fact a leakage problem may vary from zone to zone and country to country. Each water supply manager will be able to develop his/her own 'expected' indicators from which he/she will be able to identify if there is a problem in the zone or not. In this example, the expected minimum night flow was estimated to be ±20 m^3/hr, while the measured value was ±80 m^3/hr as shown. In very simplistic terms, this suggests that there is a high level of unexplained leakage of ±60 m^3/hr as shown in Figure 6.6. The calculations are never exact and the main purpose of the analysis is to establish whether or not there is a major leakage problem. If the calculation suggests that the potential leakage problem is marginal, then effort should be placed elsewhere since there will normally be zones where the results are conclusive.

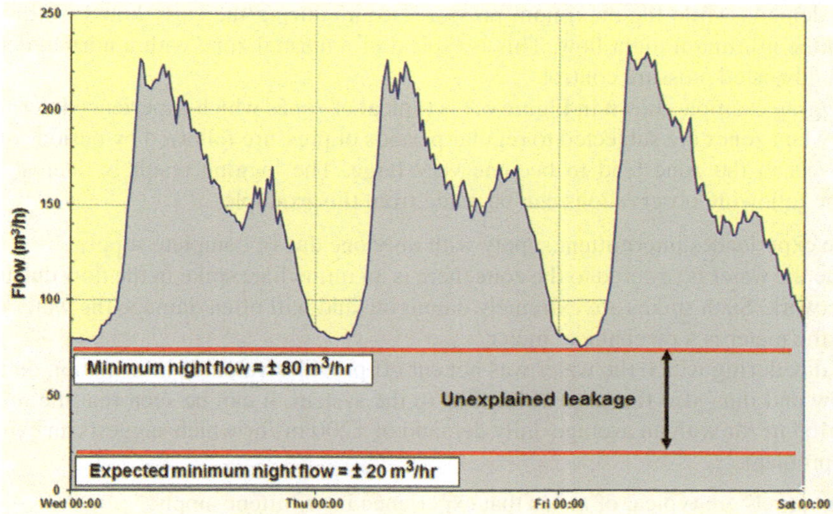

Figure 6.6 Simplified calculation of leakage.

Having investigated the flow logging results for the specific zone in question, it is then useful to look at the pressure logging results, as shown in Figure 6.7, which have been included together with the previous flow loggings.

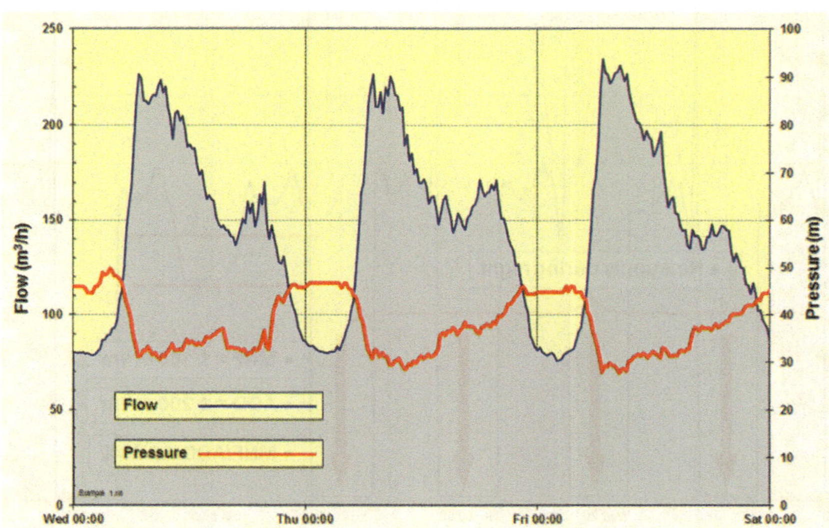

Figure 6.7 Logging with flow and pressure.

It is often very helpful to analyse the flow and pressure logging results simultaneously since they are part of the same puzzle and the flows often help to explain pressure variations and vice versa. From Figure 6.7 it can be seen that the minimum pressure of approximately 30 m occurs each morning during the main morning peak demand while the maximum pressure occurs during the early hours of the morning and corresponds to the minimum night flow. This is typical of a normal zone with a normal demand pattern and no form of advanced pressure control.

The flow logging example shown in Figure 6.8 is typical of zones which experience intermittent supply. Unfortunately when zones are subjected to regular periods of pressure followed by periods of no pressure, the leakage levels in the zone tend to become very large. The logging result is interesting in several respects and the following observations can be made from this example:

- The zone experiences intermittent supply with only one day of complete supply
- Each time the water is restored to the zone there is an immediate spike in the flow due to the refilling of the network. Such spikes are extremely damaging and will often damage the water meter beyond repair if the meter is a mechanical meter.
- The one day during which the water was not cut off provides valuable information on the minimum night flow and thus also the level of leakage in the system. It can be seen that the minimum night flow is ±165 m^3/hr with an average daily demand of ±200 m^3/hr which suggests an extremely severe leakage problem.

Such leakage levels are typical of zones that experience intermittent supply.

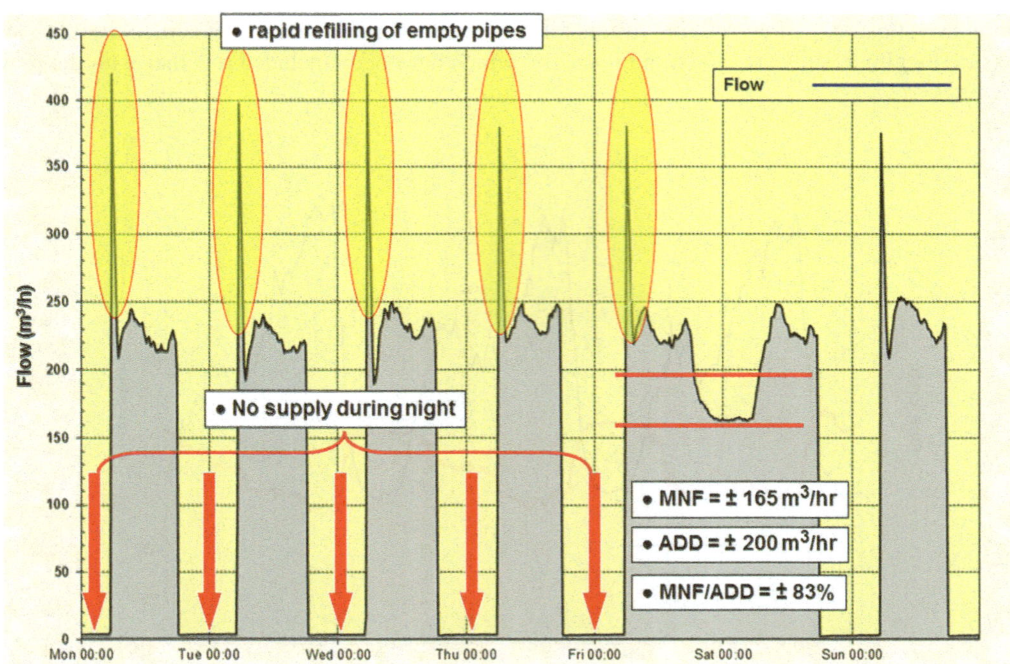

Figure 6.8 Zone with intermittent supply.

Figure 6.9 is a very interesting logging result as it highlights one of the most common problems experienced in water reticulation systems. It highlights the flow and pressure results that can be expected when a zone cross boundary valve is opened by mistake (sometimes deliberately). The logging result shows the pressure at a pressure reducing valve (red) as well as the pressure downstream of the valve as it enters the zone (purple). The flow through the PRV into the zone is shown in blue.

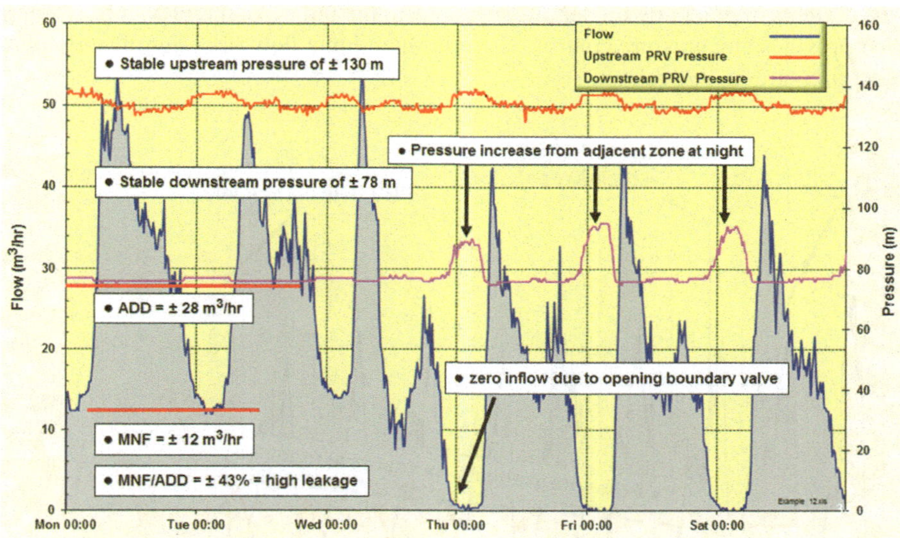

Figure 6.9 Zone compromised by open boundary valve.

When trying to understand such a complicated logging result, it is often helpful to work from left to right and start by examining the minimum night flow. If the changes to the minimum night flow can be explained, it is usually possible to explain what has happened. In this case, the minimum night flow is initially ±12 m^3/hr and after a few nights appears to drop to zero. This indicates that either all leakage, including background leakage, has been eliminated or the valve has closed and there is no flow through the valve. Since it is impossible to eliminate all leakage completely, it can be concluded that the valve has closed. On examination of the pressure downstream of the valve, it appears that there is a pressure increase each night corresponding to the zero flow through the valve. This is due to back pressure into the zone from some adjacent zone which in turn closes the PRV and results in zero minimum night flow. The previous leakage has not been eliminated and in fact will be higher during the night as a result of the increased pressure. This is a nice example to highlight what to look out for when there are problems with unauthorized opening and closing of valves within the system.

The example shown in Figure 6.10 is one of the more complicated and interesting logging results which was recorded by the City of Johannesburg at one of their PRV installations. Once again it is best understood by starting at the left and working towards the right and attempting to explain the changes to the minimum night flow. The story in this case reads as follows:

- Zone initially experiences relatively high night flow of ±40 m^3/hr
- Night flow jumps by ±20 m^3/hr on day 2 due to a mains burst which pushes the minimum night flow up to ±60 m^3/hr

- The leak is repaired during day 3 as indicated by a refilling spike and the drop in minimum night flow back to ±40 m^3/hr
- Over the next 6 nights, the PRV starts to experience problems and is unable to maintain a fixed pressure. During the day, the pressure appears stable due to the higher demand;
- The minimum night flow also gradually increases each night in response to the higher pressures. This is an excellent example to highlight the fact that pressure drives leakage.
- After the 6 nights of gradually higher pressures due to the failure of the PRV, it fails completely and no longer provides any pressure control. The minimum night flow rises significantly in response to the higher pressure.

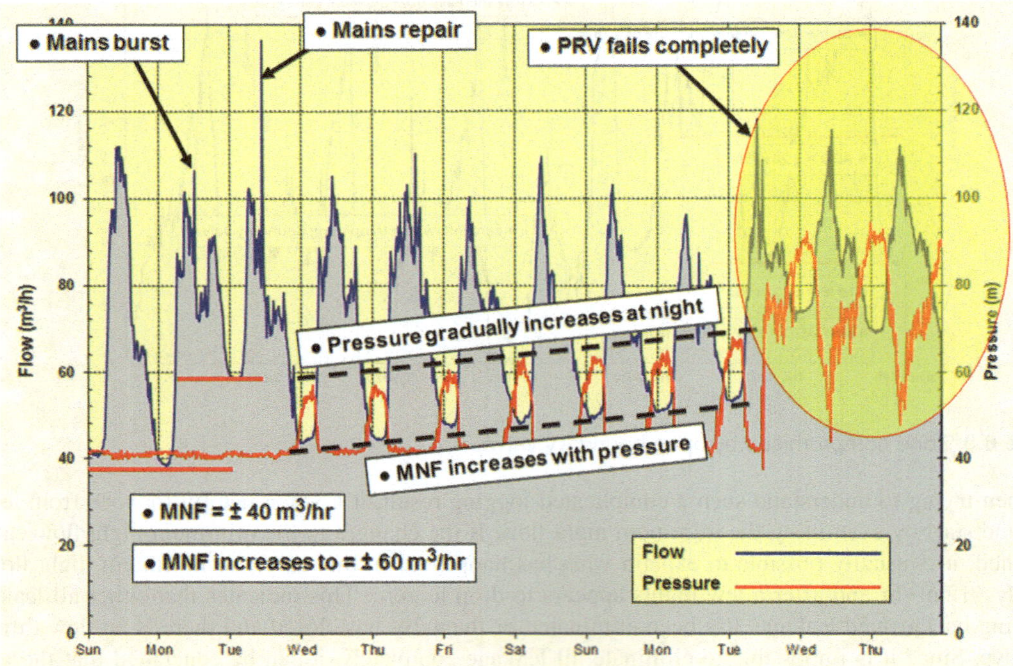

Figure 6.10 Failure of pressure reducing valve.

Figure 6.11 provides a flow logging result taken directly from the City of Tshwane's (Pretoria, SA) real time logging system and it shows the flow recorded at an installation containing both a meter and a PRV. The flow at the meter indicates that a serious leak of ±30 m^3/hr developed between the 14 and 15 of December. This leak was identified by the municipality within a day and it took a further 3 days to find and repair it. It was an interesting leak in the respect that it occurred in an area of open ground where a construction vehicle had accidently knocked over a hydrant which was leaking into an open field. No damage had been caused and there was no evidence of this leak from the nearby road as the water was flowing directly into a small stream and caused no damage or inconvenience. Without the monitoring system in place, this leak would not have been identified for weeks if not months. After finding the source of the leak, it was quickly repaired and the MNF returned to the levels prior to 14 December.

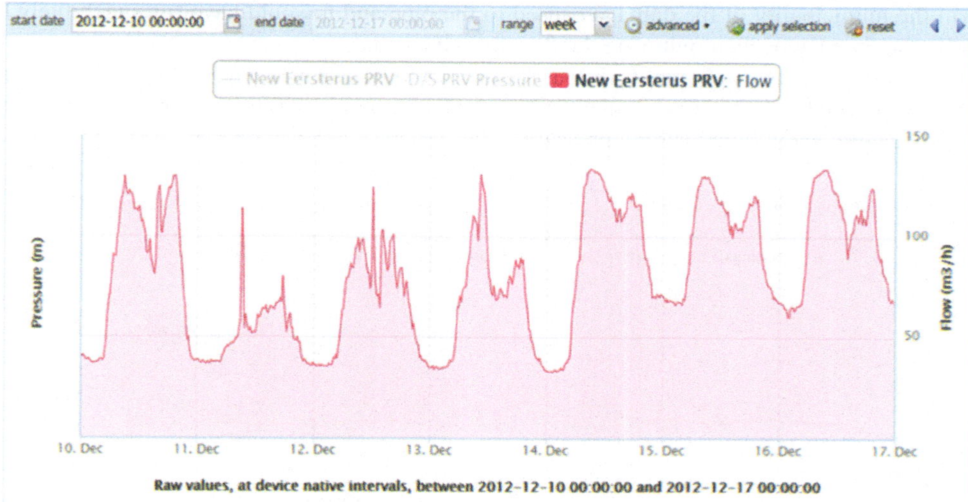

Figure 6.11 Example of mains leak from City of Tshwane (courtesy City of Tshwane).

The example in Figure 6.12 shows a leak of 40 m³/hr that developed between 4 and 5 April. In this case, the leak was identified from the real-time logging system and it took almost a week to find and complete the repair. It is important to appreciate that not all leaks create problems or are immediately visible.

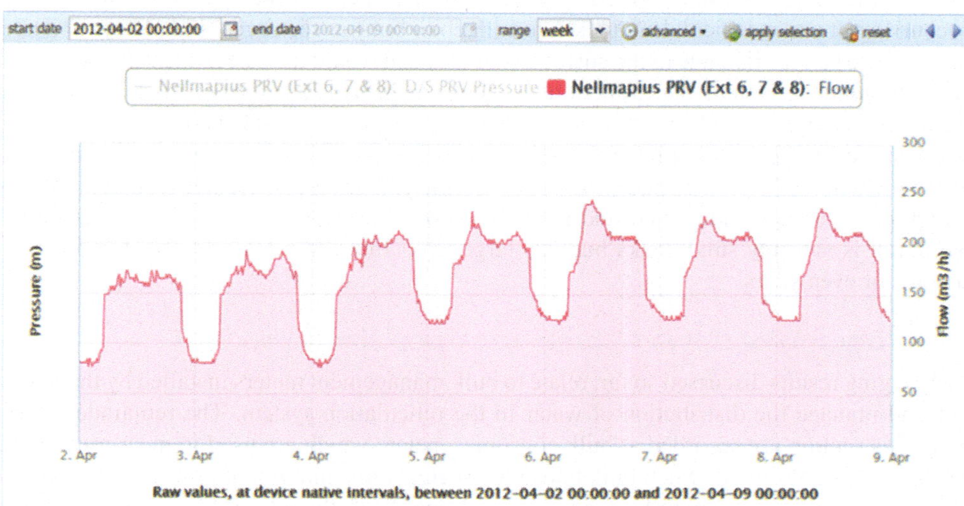

Figure 6.12 Second mains leak from City of Tshwane (courtesy City of Tshwane).

This specific leak was caused by the end cap on a 150 mm diameter pipe being damaged by a contractor once again in an area of open field where the water from the leak ran into a stream. The contractor did not inform the municipality of the damage caused to the pipe and it would have run indefinitely had the municipality not been actively monitoring the flow into the area.

The third example, shown in Figure 6.13, was identified through an alarm from the logger as soon as the leak occurred. As can be seen, this was an enormous leak of around 1000 m³/hr from a 300 mm diameter pipe which had completely failed at high pressure.

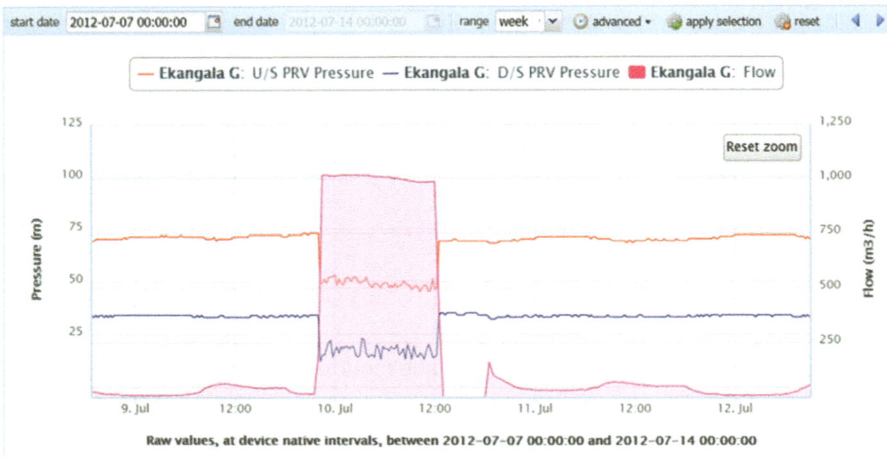

Figure 6.13 Example 3 from City of Tshwane (courtesy City of Tshwane).

Managers at the City of Tshwane were in contact with the satellite office personnel within minutes of the leak occurring. It was a difficult leak to isolate and repair but was repaired within hours of it occurring. This leak did cause serious damage to the surrounding area and would have been identified with or without the real-time monitoring system. The system does, however, provide a record of the leak from which the volume of water lost and time taken to find and repair the leak can be established.

The graph shown in Figure 6.14 highlights a leak that had been running for some time and was repaired on 7 July after which the MNF dropped by about 50 m³/hr. This is a very sizeable leak but it had not been reported and was eventually identified and repaired by Tshwane Water Department. It is important to appreciate that it is not only small leaks but also large leaks that can remain undetected for many days, weeks, months or even years.

6.1.9 Live logging of flows at bulk consumer meters

Most of the logging results discussed so far, relate to bulk management meters installed by the municipality to monitor and manage the distribution of water in the reticulation system. The remainder of examples discussed in this section are recorded at bulk consumer meters which are used to measure water used by individual industrial consumers. Such bulk consumer meters are generally not used to analyse minimum night flows but are rather used mainly to generate revenue. With the advent of new logging technology and the internet, it is now not only possible but in many cases, highly cost effective to log and analyse the water being supplied to a single large consumer. Such consumers are often the main source of income to the municipality and therefore it is very important to ensure that all water used by the consumer is properly metered and billed. From the customer's viewpoint, it is also important that they can identify any wastage and leakage on their network since the costs involved can often be significant.

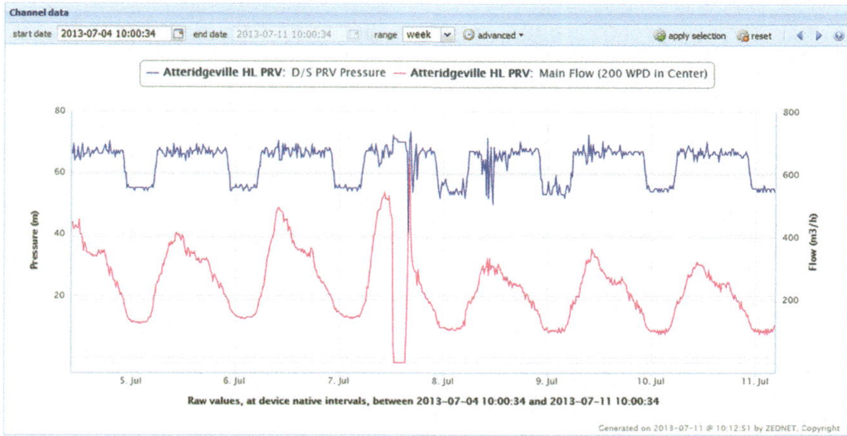

Figure 6.14 Leak and repair in Attridgeville (courtesy City of Tshwane).

The example shown in Figure 6.15 is an interesting case in which a small industrial consumer was found to have a buried connection with a functional meter which did not appear on the billing system. When the connection was discovered through the use of metal detectors, it was found that the meter was running continuously suggesting the presence of a leak. The meter was logged to assess the extent of the leak and the results of the logging exercise are as shown in Figure 6.15. As can be seen, the leak appeared to have been running at ±24 m^3/hr, which is a very substantial leak. Under normal circumstances, such a leak would often create some evidence on the surface in the form of a puddle or a damp area. In this case, however, it was found that the leak was running directly into a sewer and was causing no damage or visible signs of dampness. After the leak was repaired, the water supply through the meter dropped dramatically, as can be seen in Figure 6.16.

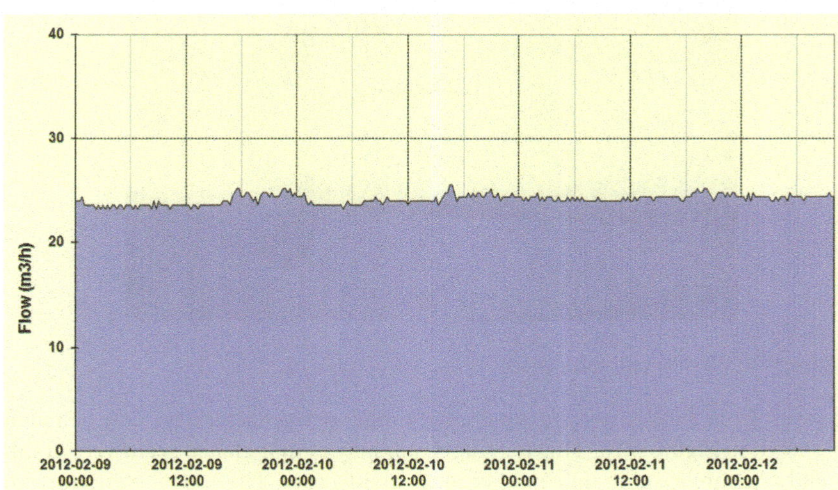

Figure 6.15 Industrial flow logging before leak repair.

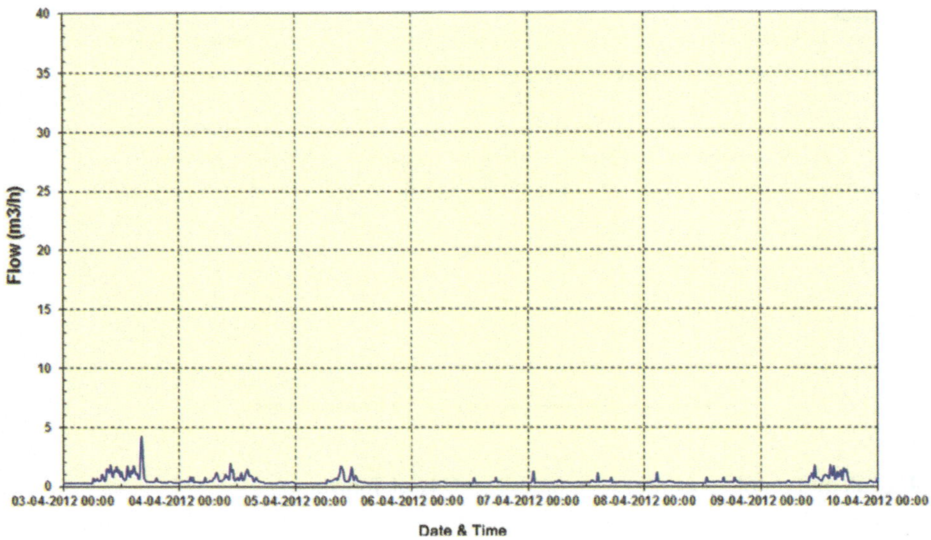

Figure 6.16 Industrial logging after leak repair.

The damaged pipe that created this particular leak is shown in Figure 6.17, which indicates that it was a split in a 63 mm uPVC pipeline. Based on the meter reading, it was estimated that this leak ran for over 3 years and cost the municipality ±$300 000 based on their bulk water purchase cost.

Figure 6.17 Damage to 63 mm diameter pipe.

The logging result from another project where real time logging has proved very beneficial and highly cost effective is shown in Figure 6.18. This logging result may at first glance seem very complicated but is in fact the flow and pressure logging into a major industrial factory. The results, which are shown in more detail in Figure 6.19, clearly highlight the impact of a pump being switched on and off throughout the day which in turn creates large pressure variations throughout the network in the vicinity of the factory. Such large fluctuations should be discouraged and the industries concerned should introduce some form of on-site storage to attenuate the variations in demand and pressure.

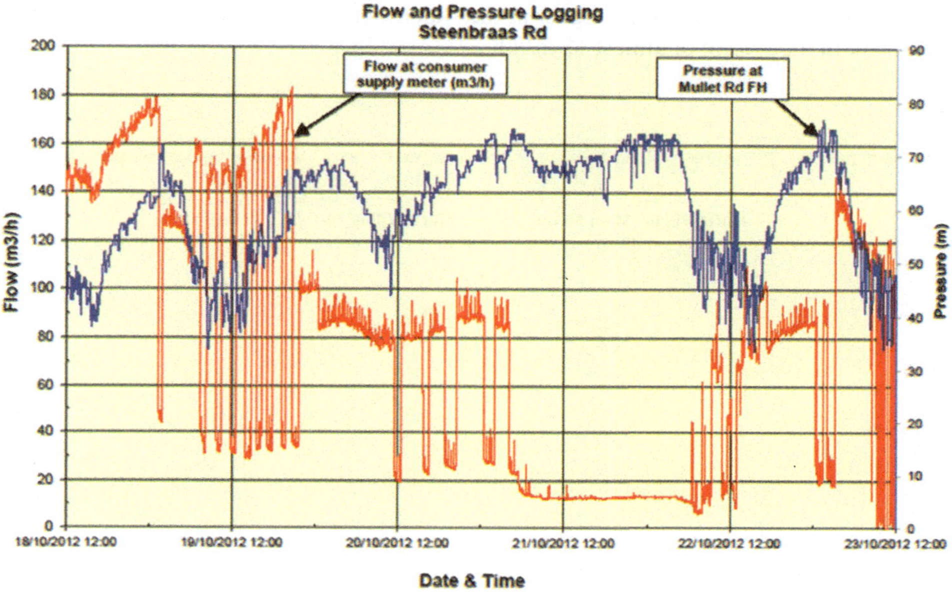

Figure 6.18 Logging result from Wadeville Project in Ekurhuleni.

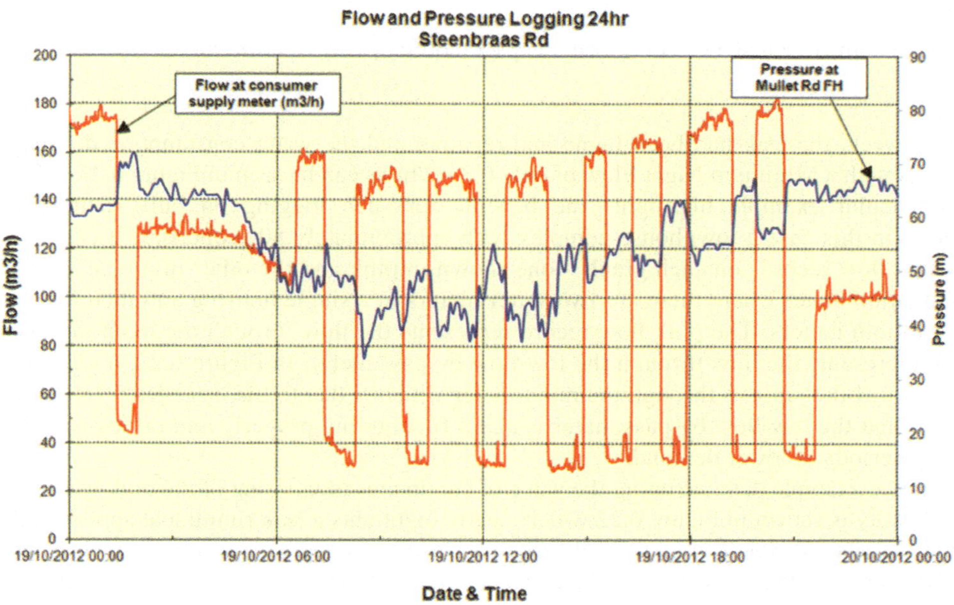

Figure 6.19 More detailed examination of industrial water consumption.

The graph shown in Figure 6.20 provides an example of a relatively small leak which had been running inside a town-house complex for many months. The residents of the complex had approached the Municipality for assistance regarding their monthly water bill which had been slowly increasing over a period of two years. After installing the logger at the water meter to the complex, it was clear that there was a serious leak in the system of almost 15 m^3/hr.

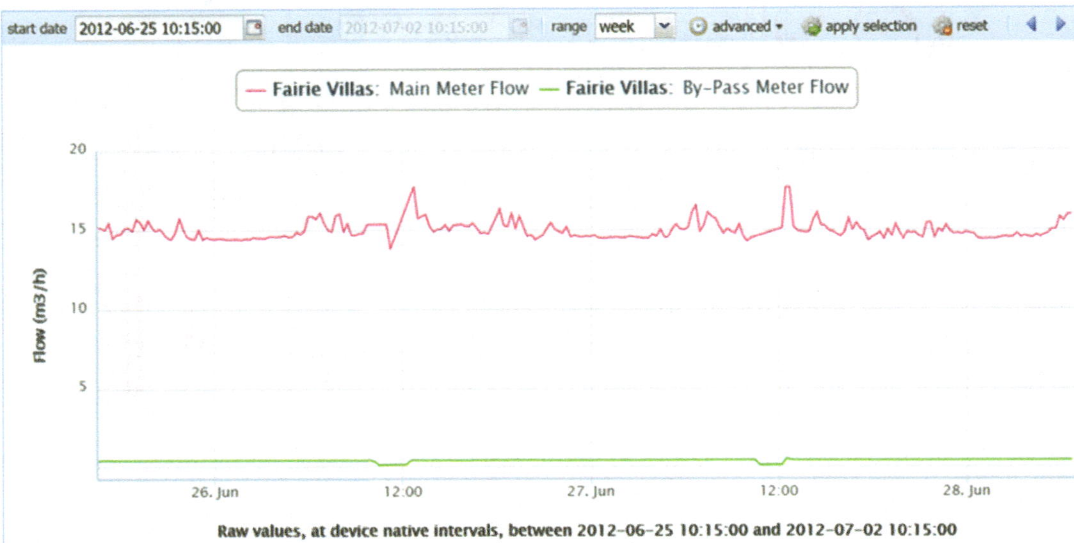

Figure 6.20 Example of leak in town-house complex (courtesy City of Tshwane).

Within a week, the leak had been found and repaired and the water consumption dropped back to normal levels with a Minimum Night Flow of only 0.3 m^3/hr as can be seen in Figure 6.21.

This particular example highlights the benefits for such logging on bulk meters into large consumers – in this case a townhouse complex with approximately 50 properties. It can be seen that there are two flow records on each graph – one shown in pink and the other in green. This is typical of a combination meter where there are two water meters in a single housing and therefore the logger must record both meters. The pink flow record represents the flow through the main meter while the green line represents the flow through the low-flow by-pass meter. In Figure 6.20, it can be seen that virtually the full flow passes through the main meter which is due to the large leak. In Figure 6.21, it can be seen that the low-flow by-pass mater is again functioning properly and taking up much of the flow during periods of lower demand.

Another nice example demonstrating the value of continuous monitoring of the bulk water use into an industrial facility is shown in Figure 6.22 which clearly highlights a leak running at approximately 2 m^3/hr. Without continuous monitoring, the industrial user would simply have received an abnormally large water account and would not have realized the extent of the leak, which is quite clear from the graph. This leak represents a loss or rather an additional expense of almost $2000 per month to the industrial user based on the current industrial water tariff in the area.

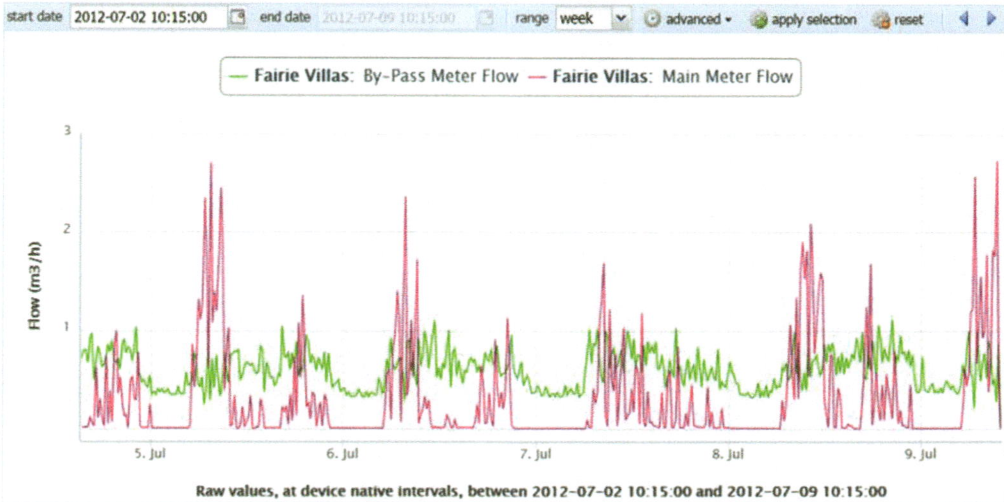

Figure 6.21 Post-repair graph from Example 4 (courtesy City of Tshwane).

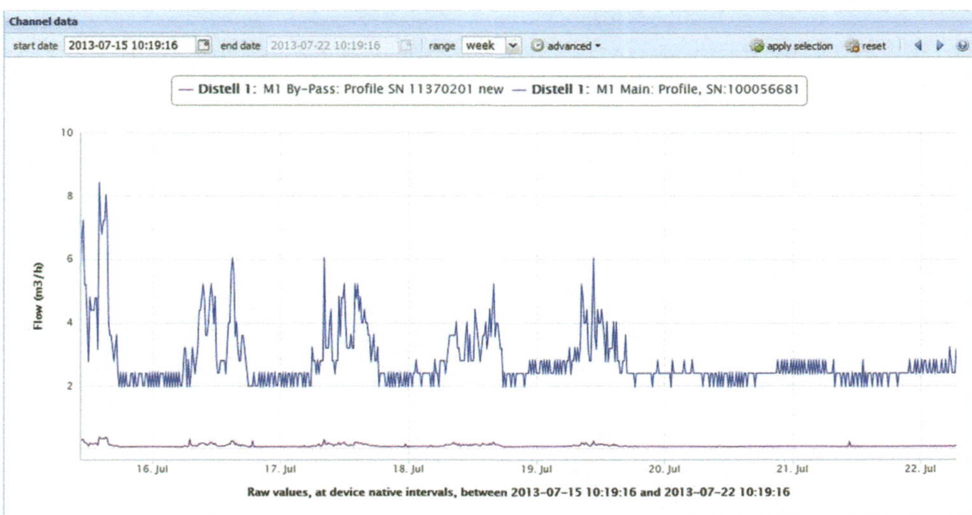

Figure 6.22 Bulk consumer leak (courtesy Ekurhuleni Metro).

REFERENCES

Mckenzie R. (2014). Remote flow logging as tool to drive down leakage. International Water Association, *Water Loss 2014*, Specialist Conference, Vienna, March.

WRC (1999). Development of a Standardised Approach to Evaluate Bursts and Background Losses in Water Distribution Systems in South Africa: SANFLOW. Report TT109/99, by R Mckenzie to the South African Water Research Commission, June, ISBN No. 1 86845 490 8.

Chapter 7

Large consumer meter consolidation and improvements: Planning, implementation and benefits

7.1 INTRODUCTION

The efficient metering and billing of large consumers is of great significance to the revenue stream in any municipality. In the case of heavily industrialized municipalities, large industrial water users can often account for more than 50% of the total water supplied and it is therefore essential that this water is accurately metered and billed.

Past practices of allowing un-metered fire connections to large consumers has resulted in some abuse of the un-metered connections and in many cases, the fire connections have simply been ignored and over a period of years, have effectively been lost. Some examples of the meters as they are found in the ground are shown in Figure 7.1 and Figure 7.2.

Figure 7.1 Broken consumer meter.

Figure 7.2 Below ground meter chamber.

In municipalities which have significant industrial water use, the accurate metering and billing of the industrial consumers is often one of the first and most important WDM interventions that should be considered. Industrial consumers are also generally willing to pay for all water used, unlike some of the residential consumers who can be problematic in some countries where payment levels are low. The remainder of this section will describe the approach and general methodology that has been applied in one heavily industrialized municipality in which a pilot project involving over 180 large industrial customers was considered. The success of the project has since resulted in a full-scale project to investigate a further 25,000 large industries in the municipality.

The water infrastructure supplying many of the industries had not been checked or upgraded for many decades, during which the municipal records in many cases had become corrupt and key billing information had been lost. As a result of these problems, the municipality was losing a large portion of its potential income from the industrial customers who are generally willing and able to pay for the water they use. Unlike the situation with the domestic customers, most of the industrial customers wish to know how much water they are using and are willing to pay for a reliable and well managed supply. The problem facing the municipality in this example with regard to the industrial water consumers was compounded by the fact that many of the properties had multiple water connections which created significant challenges to the metering and billing systems.

In order to address the metering and billing problems, it was decided to undertake a full and comprehensive meter consolidation project (Astrup *et al.*, 2013; Astrup, 2014). The aim of this project was to ensure that every supply to a specific industry was identified and where possible consolidated into a single supply point which could be accurately metered. This objective could only be achieved through the application of a zero pressure test after a consolidated supply through a single connection was designed and commissioned.

7.1.1 Planning and policies

Municipalities must make policy decisions which will define how such projects are implemented. These decisions relate to, among others, the equipment which will be used, how meter installations will be constructed and the method in which multiple connections will be consolidated. Some of the issues to be considered are as follows:

7.1.1.1 Single or multiple connections per property

Reducing multiple connections to a single connection per property reduces downstream management and maintenance costs. It is recommended that wherever possible multiple connections be consolidated. The consolidation cost must also be considered when making a decision to consolidate or allow multiple connections.

Combining domestic and fire reticulations can cause a water quality problem when stagnant water in the fire system circulates back into the domestic system. This can be overcome by installing non-return valves on the fire connections downstream of domestic off-takes. In this way, fire and domestic water can be fed from a single connection.

7.1.1.2 Standard meter design

A standard meter design is recommended to manage quality and have consistency within a water supply utility. The standard design should detail whether the meter is installed inside or outside the consumer property; above or below ground; with or without a strainer; with or without a non-return valve (NRV) and the position of isolating valves upstream and downstream of the meter.

In this specific example, it was agreed at the start of the project that all meters would be installed above ground and outside the property to aid locating and meter reading. Isolating valves were included upstream and downstream of the meter; a strainer was installed upstream of all mechanical meters and a non-return valve was installed downstream of the meter to prevent any backflow from the industry into the municipal reticulation. An example of a typical industrial water supply installation is shown in Figure 7.3.

Figure 7.3 Example of typical large consumer meter installation.

7.1.1.3 Selection of meter type

As non-domestic water consumers typically have a large fluctuation in water demand, a meter with a high turndown ratio is required to accurately meter the high and low flows. Combination meters with a large accurate measuring range have been used for many years to accommodate this problem since they direct flow through a small bypass meter during low flow conditions and have a change-over valve which opens to allow flow to pass through the main meter during higher flow conditions. Although such combination meters do address the large metering range, they often experience other problems such as sub-meter failure and the changeover valves becoming stuck.

In the past few years, new generation single element mechanical meters, with low flow measurement capabilities, are able to compete with combination meters at low flow rates and do not have the problems associated with combination meters. Such meters are worthy of consideration in most cases when selecting a meter type.

Magnetic flow meters are also appropriate in most cases and particularly for very large water users where low flow rates are seldom experienced and high accuracy is of particular significance. One of the main advantages of a magnetic flow meter is that the flow measurement accuracy of the meter does not deteriorate with time, which is a problem with mechanical meters. As with all types of meter, the magnetic flow meters do have some disadvantages, one of which is the cost since they tend to be more expensive than the mechanical meters and in some instances can be damaged by lightning. Some magnetic flow meters are battery powered in which case a mains power supply is not required. In cases where mains power is required, this can be problematic, especially in areas where cable theft is an issue.

Typical meters used for metering non-domestic consumers are illustrated in Figures 7.4, 7.5 and 7.6.

Figure 7.4 Sensus Meitwin combination meter.

Figure 7.5 Elster Kent H5000 full flow mechanical meter.

Figure 7.6 Krone Waterflux Magflow meter.

7.1.1.4 Revise policy on fire connections

Unmetered connections to consumers provide a potential for unaccounted water loss from the municipal water system and should be eradicated. Water supply authorities should therefore review and if necessary revise their policies on unmetered fire connections. Unless strong policing of unmetered fire connections exists, all fire connections should be metered.

7.1.1.5 Consolidate internally or externally

The consolidation of multiple water connections to a property will generally involve the installation of a single meter from the council main and connection of all of the existing supplies to the property downstream of this meter. It is recommended that pipework required to consolidate multiple water supplies be installed inside the consumer property. This way, the consumer will own and be responsible for the maintenance of this water infrastructure.

7.1.1.6 Evaluate hydraulic capabilities of the municipal network

It is important to determine the fire fighting capabilities of the municipal water supply network. This can be achieved by opening a fire hydrant fully and monitoring the flow delivered by the fire hydrant and the impact that this flow has on the pressures in the municipal network. It is impractical for a consumer to have a large fire-fighting connection when the municipal system is not capable of delivering sufficient flow to match the connection size. In cases where the municipal network is not capable of delivering sufficient flow for fire-fighting purposes, the consumer should be informed of such limitations and advised to erect on-site storage to accommodate their specific fire-fighting requirements.

7.1.1.7 Develop a workflow methodology

Developing a standard workflow methodology is critical to successfully undertaking and managing such projects. An example of a typical workflow methodology is illustrated in Figure 7.7.

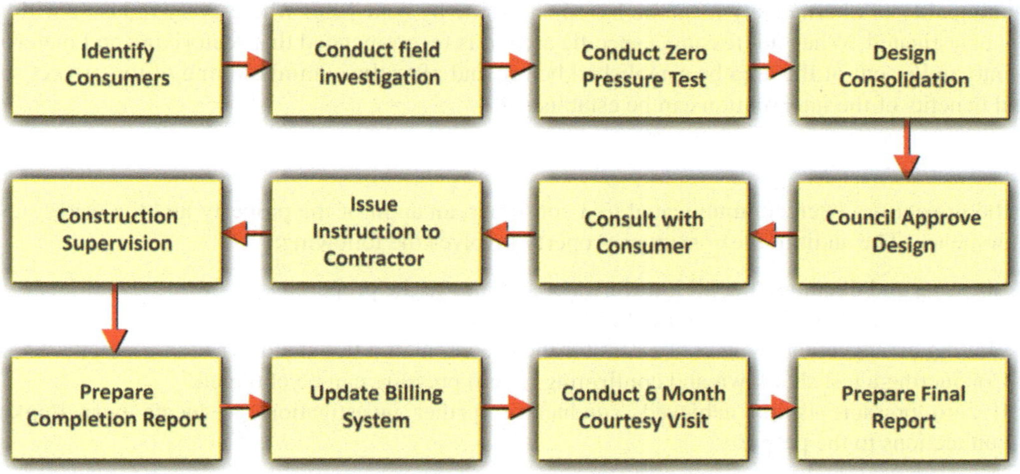

Figure 7.7 Typical workflow tasks for meter consolidation and improvement project.

7.1.1.8 Standard documentation & templates

In order to properly manage and monitor the work involved with several hundred large industrial meter consolidations, standard documentation and templates are required. Effective templates greatly improve the quality of work and reduce the time taken to execute the project.

7.1.1.9 Meter sizing

In order to consolidate multiple meters into a single supply it is necessary to carry out a meter-sizing calculation for each new installation. A standard form for capturing all water use requirements on the property such as fire, process/commercial and domestic water use is required and should be used to calculate the appropriate meter size for the property. The capability of the municipal system to deliver flow should also be considered when sizing the meter installation.

7.1.1.10 Zero pressure test

A zero pressure test is conducted by isolating all supplies to a property and confirming zero pressure on the property. These tests are used to identify the presence of any unknown connections and ultimately eradicate these. There is a cost and inconvenience associated with conducting such tests. The long-term benefits however outweigh the initial cost and inconvenience. A policy decision needs to be taken as to whether such tests will be undertaken at each and every property or not. Every test must be properly witnessed and signed off by a representative of the municipality on a standard form.

7.1.2 Implementation

7.1.2.1 Identify areas/consumers for implementation

Before proceeding with any meter consolidation project, it is important to identify which industrial consumers will be investigated. This can be achieved by selecting the top 100, 200, or 500 consumers in the municipality or alternatively it can be based on a zone by zone basis where every industry within the zone is investigated. When addressing a specific area, it is recommended that sectorizing and metering of flows into and/or out of the area be established before and after the commissioning of the project so that the real benefits of the intervention can be established.

7.1.2.2 Consumer audit

Once the project has been communicated to a consumer, an audit of the property and the water supplies can commence. The audit of the consumer property involves the following:

- Locating and documenting all known water supplies to the property.
- Investigating and documenting the fire-fighting infrastructure on the property.
- Making arrangements with consumers for the execution of a test shutdown.
- Conducting a test shutdown and confirming if zero pressure can be obtained.
- If zero pressure is not achieved, conducting further investigations to locate buried/unknown connections to the property.

An aerial view illustration of connections located for a typical large consumer is illustrated in Figure 7.8.

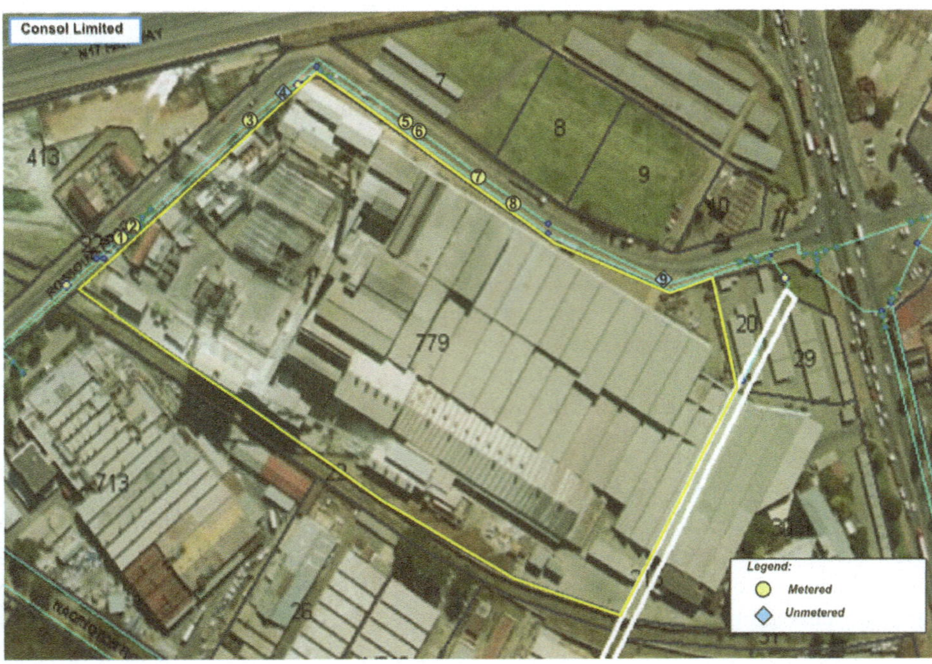

Figure 7.8 Example of metered and un-metered supplies to a large consumer.

When undertaking the assessment, the following information should be collected for each connection/supply point:

- meter size;
- meter make and model;

- meter co-ordinates;
- meter serial number;
- meter reading and date of reading.
- photographs of the meter.

A typical audit form is included in Figure 7.9.

Meter (1)			
Meter Supplying:	*Fire and Production*	Serial Number:	*XZJ144*
Supply Size up-stream:	*150*	Factor:	*1*
Supply Size dn-stream:	*150*	Reading (m^3 only):	*1319099*
Meter Size (mm):	*150*	Legible (y/n):	*Yes*
Make:	*Kent*	Working (y/n):	*Yes*
Model:	*Helix 3000*	Meter Position:	*OutsideBelowGround*
GPS Latitude (S):	*26.25734*	Is a Sub Meter (y/n):	*No*
GPS Longitude (E):	*28.18526*	Parent Serial Number:	
Date of Manufacture:		:	
Comments & Recommendations			
Photos of Meter Installation			

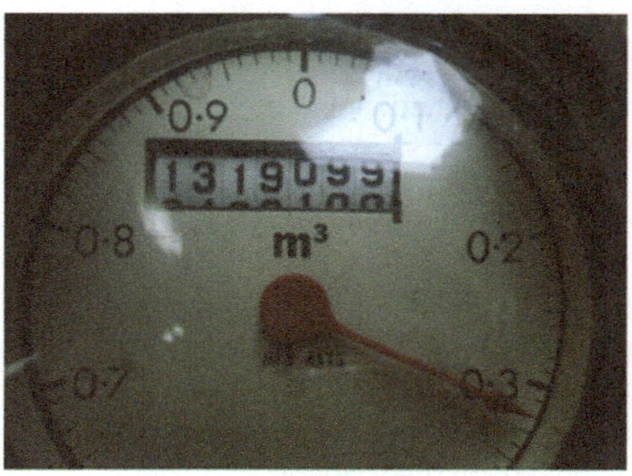

Figure 7.9 Illustration of information collected for each consumer meter.

7.1.2.3 Design

Once the audit has been completed, the design phase of the meter consolidation can commence. The design report should include all relevant information collected during the audit phase and detail exactly how the existing supplies to the property should be consolidated or upgraded. The consumer should be consulted during the design phase on any possible problems regarding their meter sizing or supply consolidations.

The design report should provide the following:

- Sizing of the new meter;
- Schematic depicting how supplies will be consolidated including comments on the consumers existing water account.
- A letter informing the consumer of the required changes to their water infrastructure.

The design report should be submitted to the municipality for approval prior to issuing to a contractor for construction.

7.1.2.4 Construction

Once the design has been approved by the municipality, negotiations with the consumer to gain approval of the design can commence. Thereafter the contractor can be issued an instruction to proceed. Close supervision and involvement by the supervising engineer is recommended during the construction phase to ensure quality and to capture all relevant meter information.

A major challenge with work on existing water supply systems is the ability to isolate sections of the network. Careful review of existing drawings and location and verification of existing valve infrastructure is required. This task is critical to the success of the construction phase. Support of municipal operations staff greatly improves the ability of the teams to find the required infrastructure.

Details of all old and new meter readings and serial numbers must be photographed and properly documented during the construction phase. Open communications with the consumer and other affected consumers and the contractor must be maintained throughout construction.

7.1.2.5 Capturing information

Once the connections have been consolidated, it is imperative that all old and new meter information is correctly uploaded onto the municipal billing system. Every meter removed must be closed on a final reading (and date) and all new meters installed must be captured onto the municipal billing system. The consumer accounts must be uploaded with final meter reading information and corrections must be made where interim readings have been applied due to lost or missing meters.

7.1.2.6 Six month courtesy visit

A six-month courtesy visit is recommended in order to evaluate the status of metering on the property six months after consolidation and discuss any concerns that the consumer may have. The zero pressure test should be repeated to ensure no unknown connections to the property and the meter should be inspected for any vandalism. In addition, the strainer should be inspected and cleaned. The consumer account should be reviewed to ensure that the correct meter information appears on the account.

7.1.2.7 Generate full report

A full report is generated to consolidate all the information relating to each bulk consumer consolidation into a single document. This report should include all of the previous reports such as the design and completion report as annexures. The report should also include information collected during the courtesy visit. All meter reading information for the consumer should be summarized in the report in the form of tables for easy reference. The report should provide details of the benefits achieved through the intervention such as the increase in metered consumption on the consumer property following the consolidation intervention.

7.1.2.8 Development of a management database

Although a substantial amount of work is required to successfully consolidate each consumer, the work is highly repetitive for each consumer consolidation. Sub-tasks are repeated for each phase of work and a large volume of data is collected. This provides an ideal opportunity for the development of a structured database to facilitate the collection, interrogation and display of this data. The database can also be used to produce standard reports as required throughout the life cycle of interventions. Ideally the database should also have a facility to generate job cards which are linked to information collected through earlier phases of work for issue to a contractor.

7.1.3 Results

7.1.3.1 Reduction of commercial losses

By eradicating un-metered connections, replacing old or inoperable meters and consolidating supplies, commercial losses will be reduced. The improved metering and ease of access to these meters facilitates the accurate metering and billing of consumers. In addition, the identification and metering of buried 'lost' meters will have a significant influence on the commercial losses.

In the case of the project, a total of 182 consumer properties were successfully consolidated and the overall benefits were carefully audited by the project team and municipality. The increase, or decrease, in metered consumption to these properties was determined by calculating the average metered consumption on each property for six months prior to meter consolidation and comparing this to the average consumption on each property post-meter consolidation. Sixty-three per cent of the total increase in metered consumption was achieved at 10 of the 182 properties consolidated. Results of the ten properties with the largest increase in metered consumption are illustrated in Figure 7.10. On average, a 44% increase in metered consumption was achieved at these ten properties.

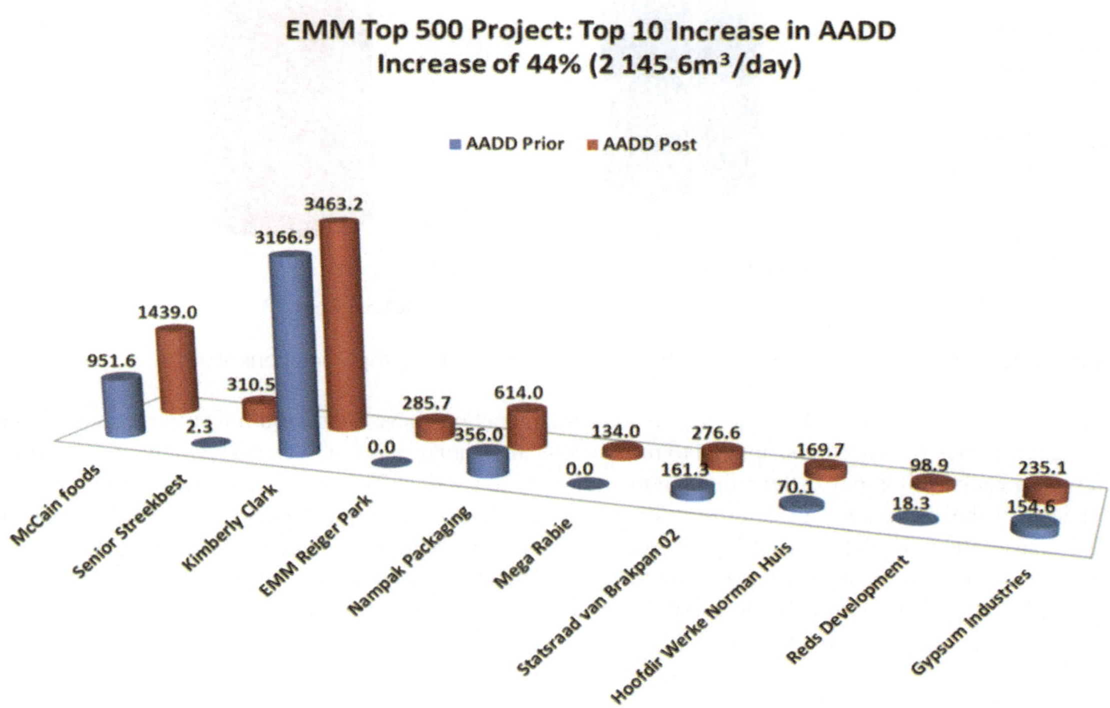

Figure 7.10 Summary of top 10 increased metered consumptions.

Seventy-two per cent of all properties consolidated experienced an increase in metered consumption with an average increase in metered consumption at these properties of 44%. Some properties experienced a decrease in metered consumption which was partly due to consumer initiatives to actively reduce water demand following the consolidation intervention and partly due to changes in commercial or production activities.

In total, the metered consumption for the area increased by 28% for all 182 properties evaluated, as illustrated in Figure 7.11. The increase in metered consumption of ±3400 m³/day, represents an increase in metered billable consumption of 1,240,000 m³/annum. This represents a revenue increase of $1.4 million/annum. As consumers are charged for sewerage based on water demand, an additional revenue increase was attained through increased sewer charges. This additional revenue amounted to ±$400,000 per annum. The gross increase in revenue collection for the 182 properties consolidated is therefore ± $1.8 million.

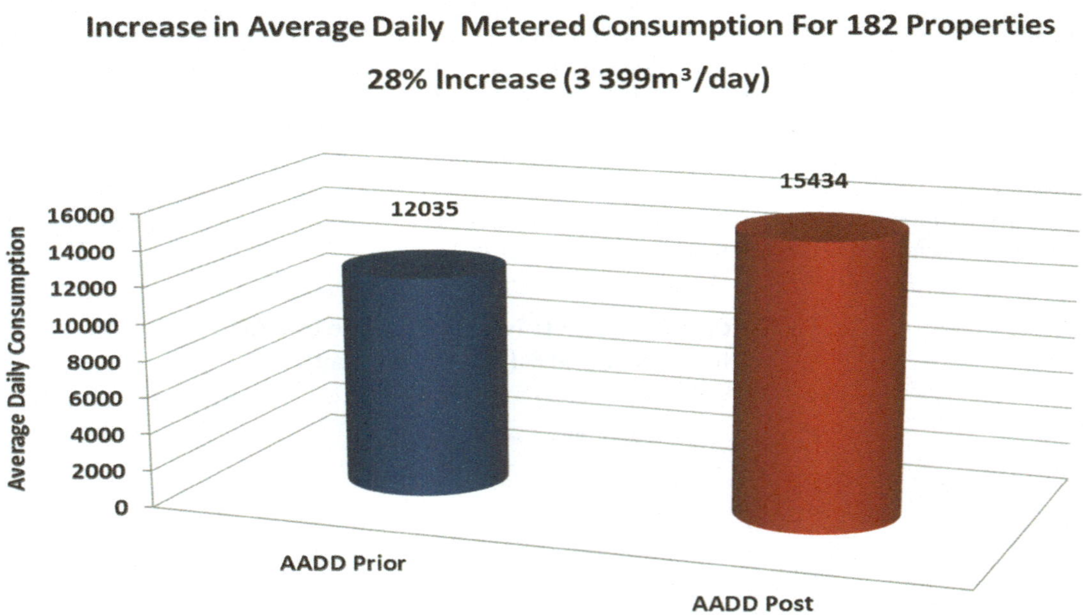

Figure 7.11 Summary of metered consumption increase for all 182 properties consolidated.

In cases where un-metered connections or illegal connections have been used for purposes other than fire-fighting, back-billing can be applied to recover this unauthorized use of water. The municipal by-laws relating to back-billing may differ from one municipality to another. The applicable water supply by-laws will need to be scrutinized to determine in which case back-billing can be applied. Some municipal by-laws also allow for the back-billing of consumers for inoperable water supply meters. Any additional revenue collection through back-billing should not been taken into account when determining the payback for the project as there is no guarantee that this back-billing will be collected.

7.1.3.2 Reduction in physical losses

In addition to the significant reduction in commercial losses, numerous municipal network leaks were noted during the implementation phase of the project, which led to a significant reduction in physical leakage.

In some cases leakage was identified on consumer properties which were not being metered or billed. In one instance, a leak which had occurred on a 63 mm uPVC pipeline adjacent to an existing sewer manhole had been running at a flow rate of 24 m^3/h for an estimated three years. Once metering had been corrected, this leakage was identified and repaired by the consumer. Such loss is considered to be physical loss to the council as it was not billed for previously.

7.1.3.3 Reduction of consumer demand

In some instances, consumers took action to reduce their demand following the consolidation and upgrading of their supply meters. This was an additional benefit of the project due to the fact that the particular municipality used in the example is situated in a highly water stressed environment, where the national water authority has mandated local authorities to reduce their water demand in order to preserve the water resources of the country.

7.1.3.4 Socio-economic benefits

Where implemented professionally and comprehensively, these interventions have greatly improved the relationship between large water users and the local authority. In general the initiative was well received by large water users.

The interventions create an excellent opportunity for job creation. A number of jobs can be created for staff from consulting engineers and contractor's alike. The specific project used in the example was spread over a three year period and created 14 full time jobs for engineering staff and 16 full time jobs for construction staff.

7.1.3.5 Project pay-back

The project was undertaken between 2009 and 2012 and involved the consolidation and improvement of meters at 182 large consumers. The cost to consolidate the 182 properties, including professional, construction and material costs, amounted to approximately $1.5 million. The intervention resulted in an increase in revenue to council of approximately $1.8 million per annum indicating a project pay-back period of less than 12 months.

7.1.4 Conclusion

Replacement of meters to large consumers in many municipalities is often a very important and cost effective WDM intervention. By applying a clear and coherent project methodology, it is possible to reduce commercial losses and improve revenue collection. The implementation strategy for such projects should include the eradication of all un-metered connections, consolidation of multiple connections into a single connection and correct sizing of consumer meters.

The accurate capturing of all removed and new meter information onto the municipal billing system is important to ensure that account information is updated. Failure to implement this correctly will result in the municipality not attaining the benefits of the physical infrastructure improvements.

In the example presented, the municipality achieved a project payback period of less than 12 months. Many water authorities throughout the world and especially in developing countries have similar large water users with poor metering infrastructure in which similar paybacks can be achieved. It is recommended that such projects be implemented in all municipalities with poor meter infrastructure to large water consumers.

Municipalities are advised to identify their highest water consumers and to tackle these first as the project payback period will be the shortest for these consumers. Based on the success of such interventions, additional areas and smaller non-domestic consumers should be identified and implemented thereafter.

REFERENCES

Astrup B. (2014). Wadeville Meter Consolidation Project. Paper presented at the *International Water Association Water Loss 2014, Specialist Conference*, Vienna, March.

Astrup B., Vorster J. and Mckenzie R. (2013). Large consumer meter consolidation and improvements: planning, implementation and benefits. *7th IWA International Conference on Efficient Use and Management of Water (Efficient, 2013)*. Paris, France, 22–25 October.

Chapter 8

Software tools to assist with water loss reduction in municipal systems

8.1 INTRODUCTION

8.1.1 Background

Within the last twenty years there has been a growing realization that the rapidly increasing water demands throughout South Africa are not sustainable. As a result of this realization, there has been a significant change of emphasis away from the traditional approach of developing new water transfer schemes to one of Integrated Resource Planning in which water conservation is often regarded as the top priority. New supply schemes are not excluded from future planning, but they will now be developed only in cases where it can be shown that the existing water resources are being used efficiently. Several recent studies have shown that major proposed augmentation schemes can be postponed by many years if the growth in demand can be trimmed by only a few per cent – a target that is certainly achievable in most systems. The savings associated with delaying a new water transfer scheme are often so large that the measures needed to achieve the delay are not only environmentally attractive but also very cost effective.

New legislation introduced by the South African government encourages more efficient water use (i.e., penalties for inefficient use) and will gradually result in stricter control of Non-Revenue Water throughout the country.

In order to support the government legislation and encourage efficient use of the available water resources in South Africa, the Water Research Commission (WRC) has initiated and supported numerous projects over the past 20 years. Although some very comprehensive and sophisticated software is available, both internationally and locally, it was often too expensive for many of the smaller municipalities. To address this problem, the WRC commissioned the development of numerous WDM models which are now freely available to any municipality. The models are designed to assist the municipalities in understanding and managing their water losses and overall levels of non-revenue water (NRW).

The various models are relatively simple and user-friendly, many of which are based on the Burst and Background Estimate (BABE) methodology which was first developed for the UK Water Industry in the early 1990s (Lambert *et al.* 1994). The BABE methodology has since been accepted and adopted in many parts of the world as it provides a simple and pragmatic approach to the very complex and often confusing problem of leakage from water distribution systems.

The BABE approach was first introduced to South Africa in 1994 through a series of courses and seminars presented countrywide by Ronnie Mckenzie and Allan Lambert (founder of BABE) at the request of the Water Research Commission. The methodology and concepts have since been widely accepted by most water suppliers throughout the country and have been incorporated to a large degree in the Code of Practice for the management of potable water in distribution systems (SABS, 1999).

The BABE concepts are both simple and pragmatic and are most effective when applied in conjunction with the following additional concepts:

- Fixed Area Variable Area Discharges (FAVAD) (May 1994);
- Unavoidable Annual Real Losses (UARL) (Lambert *et al.*, 1999) and,
- The Infrastructure Leakage Index (ILI) (Lambert *et al.*, 1999).

In the development of the BABE models for the Water Research Commission, it was initially agreed that the following four main elements of leakage management would be addressed, as shown in Figure 8.1:

- Logging and analysis of minimum night flows;
- Economics of active leakage control;
- Pressure management;
- Benchmarking of leakage and auditing of non-revenue water

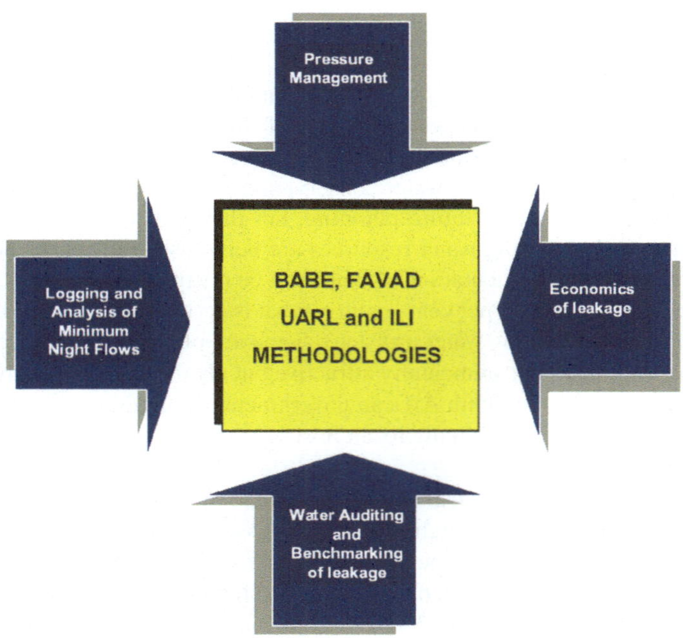

Figure 8.1 Four main elements of the BABE methodology.

The four key models shown in Figure 8.2 were therefore developed over a period of approximately four years and are regularly updated or upgraded to ensure that they remain operational under the various new operating systems. Each model is a small self-contained programme that addresses one specific issue. It was decided to adopt this simple and straightforward approach in order to avoid confusion and allow water suppliers to use one or all of the models as they consider appropriate applications rather than develop a single and more complicated application covering all aspects.

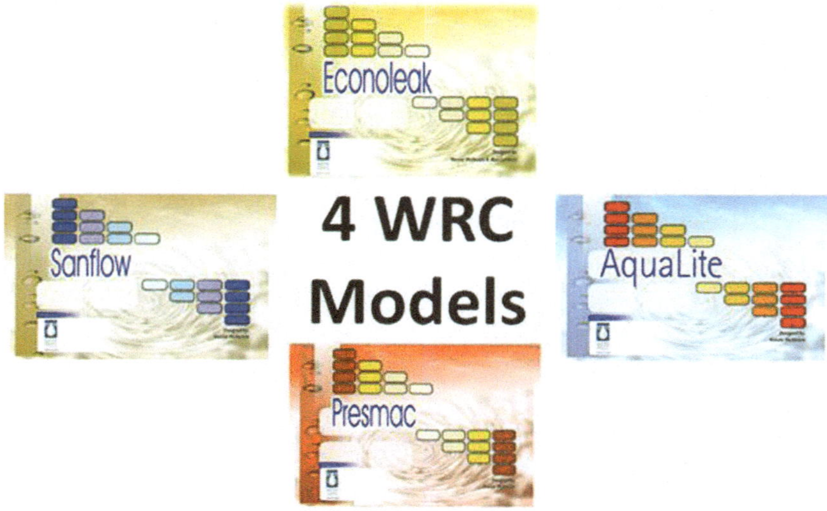

Figure 8.2 4 BABE models developed through the WRC.

The four models are freely available through the Water Research Commission and details of the models and some associated reports are provided in Table 8.1 for reference purposes. The various manuals accompanying the software can be obtained directly from the WRC website on www.wrc.org.za.

It should be noted that, while the methodologies mentioned above address certain key issues regarding the management of leakage and non-revenue water, they do not address the many social and environmental issues which are also very important. Water suppliers should therefore ensure that they consider both the social and environmental issues as well as the technical issues when implementing any new WDM intervention since the success of a project will depend on both sets of issues being adequately addressed.

8.1.2 AQUALITE: Water balance software

The AQUALITE model is basically an updated version of the original BENCHLEAK model which was one of the first Excel based water audit models to be developed using the BABE methodology. The BENCHLEAK model (WRC, 2002b) was developed from the original Word document which was prepared by Allan Lambert who assisted in the design of some of the Water Research Commission models to ensure that they were in line with the International Water Association's guidelines for 'best practice'. The BENCHLEAK model also helped to promote the concepts of the Infrastructure Leakage Index (ILI) throughout southern Africa and in turn led to a series of investigations to establish the overall level of non-revenue water throughout South African municipalities, as discussed later in this chapter.

The BENCHLEAK model was one of the first models developed by the WRC and was released in 2000, although the final documentation was only released two years later in 2002 (WRC, 2002b). The simple Excel based model was used for several years in South Africa and many other countries to assess the overall water balance of water utilities based in the standard International Water Association water balance. In 2007, the WRC commissioned the development of AQUALITE (Figure 8.3) which is effectively a Delphi based version of the original BENCHLEAK model incorporating a few additional features which were not included in the earlier Excel based model.

Table 8.1 Various WRC models and reports to assist municipalities reduce water losses.

Model	**Details**	**ISBN reference**	**WRC reference**	**Released**
SANFLOW	Model designed to provide an indication of the unexpected detectable leakage in a zone from the analysis of the minimum night flow.	1 86845 490 8	TT 109/99	1999 & 2009
PRESMAC	Model designed to estimate the potential for pressure management in a pressure zone based on logged flow and pressures over a representative 24-hour period.	1 86845 772 2	TT 152/01	2001 & 2009
BENCHLEAK	Model designed to establish the components of non-revenue water in a water utility or zone metered area and associated performance indicators based on the latest IWA Task Force recommendations.	1 86845 773 7	TT 159/01	2001 superseded
ECONOLEAK	Model to evaluate the most appropriate frequency for undertaking Active Leakage Control.	1 86845 832 6	TT 169/02	2002
AQUALITE	Updated version of model designed to establish the components of non-revenue water in a water utility or zone metered area and associated performance indicators based on the latest IWA Task Force recommendations. Replaces Benchleak.	978-1-77005-599-5	TT 315/07	2009
WDM SCORECARD	A basic scorecard based model used to develop a first-order WDM strategy for a municipality, quickly and effectively.	978-1-4312-0271-3	TT 523/12	2010
The State of Non-Revenue Water in South Africa	A comprehensive report on the state of NRW throughout South Africa based on over 130 water balance assessments using the standard IWA methodology.	978-1-4312-0263-8	TT 522/12	2013

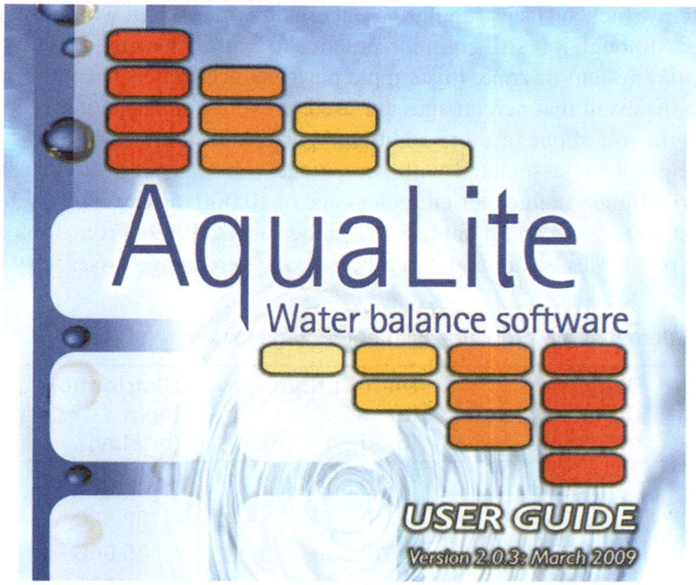

Figure 8.3 Opening screen from the Aqualite benchmarking model.

In many countries it is very difficult to compare leakage levels and levels of non-revenue water from one system to another, due to the fact that the various water suppliers adopt their own definitions for physical and commercial losses, and so on. It is only through the use of a standard and consistent water balance that figures from one water supplier can be compared to those from another water supplier. By adopting the standard International Water Association definitions, a consistent approach to the water balance can be achieved and it is then possible to compare results from different systems in a meaningful manner. The standard IWA terminology is depicted in Figure 8.4 is the basis on which the AQUALITE water audit model is based.

<table>
<tr><td rowspan="9">System Input Volume</td><td rowspan="4">Authorised Consumption</td><td rowspan="2">Billed Authorised Consumption</td><td>Billed Metered Consumption</td><td rowspan="2">Revenue Water</td></tr>
<tr><td>Billed Unmetered Consumption</td></tr>
<tr><td rowspan="2">Unbilled Authorised Consumption</td><td>Unbilled Metered Consumption</td><td rowspan="7">Non Revenue Water</td></tr>
<tr><td>Unbilled Unmetered Consumption</td></tr>
<tr><td rowspan="5">Water Losses</td><td rowspan="2">Apparent Losses</td><td>Unauthorised Consumption</td></tr>
<tr><td>Customer Meter Inaccuracies</td></tr>
<tr><td rowspan="3">Real Losses</td><td>Leakage on Transmission and Distribution Mains</td></tr>
<tr><td>Leakage and Overflows at Storage Tanks</td></tr>
<tr><td>Leakage on Service Connections up to point of Customer Meter</td></tr>
</table>

Figure 8.4 Standard IWA water balance.

One specific problem which surfaces regularly concerns the manner in which water suppliers express their levels of leakage. Although it is still common practice to express leakage as a percentage of the water supplied into a particular system or zone, this simple performance indicator can be very misleading and should not be used. In the event that percentages are used to express leakage, it is recommended that they are supported through the use of one or more additional performance indicators.

To demonstrate the problems associated with percentage values, a very simple example can be used. In this example a distribution system experiences leakage of 10,000 m³/day. This system is analysed for a range of different consumers as shown in Table 8.2 (Lambert *et al.*, 1999). From Table 8.2 it can be clearly seen that although the real losses are identical in all cases, the percentage losses vary considerably.

Table 8.2 Example to demonstrate problems with percentage losses

Per capita consumption (litres/head/day)	Daily Consumption (m³/day)	Distribution losses (m³/day)	Distribution input (m³/day)	Percentage losses
25 (Standpipe)	6250	10,000	16,250	62
50 (Jordan)	12,500	10,000	22,500	44
100 (Czech Rep)	25,000	10,000	35,000	27
150 (UK, France)	37,500	10,000	47,500	21
300 (Japan)	75,000	10,000	85,000	12
400 (USA)	100,000	10,000	110,000	9

Full details of the IWA water auditing procedure are available from numerous papers which have been presented over the past 20 years at various international conferences. In summary, however, the basic approach includes the use of the ILI which is a simple ratio of the current annual real losses (CARL) divided by the unavoidable annual real losses (UARL) and is fully explained in the original article by Lambert *et al.* (1999).

ILI = CARL/UARL

The unavoidable annual real losses (UARL) can be easily assessed for most systems as long as the number of connections, length of mains, location of customer meters and average operating pressure are known. Further details of the calculations are provided in the original BENCHLEAK User Guide (WRC, 2002b) which is available from the WRC.

Following the development of the BENCHLEAK and AQUALITE models, they have been used to assess the level of non-revenue water throughout South Africa through a series of assessments which have been completed bi-annually since 2007. It is interesting to note that the ILI values for the South African systems range from 1.0 to approximately 20, with an average value in the order of 7. This can be compared to ILI values calculated by the IWA Task Force for 27 supply systems in 19 countries which range from 1.0 to 10 with an average value of 4.2. For South African conditions it would be unusual to achieve an ILI value of below 2.0. Values in the order of 5.0 are relatively common and represent systems in a reasonable condition.

While there is still some debate concerning the use of the ILI as a key performance indicator by certain water conservation specialists, the fact remains that it provides a useful system specific indication of the leakage problem in an area in a manner which can be easily understood, particularly in developing countries. While it may not be completely foolproof, it has been used with considerable success in identifying key problem areas in large reticulation systems and as far as the clients are concerned that is exactly what they wish to achieve. At the end of the day, the results speak for themselves and until there is an acceptable alternative, the ILI will be used to assist water suppliers to manage their leakage problems.

8.1.3 SANFLOW: Background night flow analysis model

The analysis of minimum night flows is considered to be one of the fundamental issues that can assist water supply managers in assessing and managing water losses from municipal water supply systems. If used properly, the minimum night flow is one of the most useful and reliable indicators of leakage in a system. To assist water supply managers in understanding and assessing their leakage, the SANFLOW model (WRC, 1999) was developed (Figure 8.5). The model is used to assess the level of excess leakage in a discrete zone from the analysis of the minimum night flow using the standard BABE methodology, which is fully described in the accompanying documentation.

Figure 8.5 Opening screen from the SANFLOW night flow analysis model.

This model has been one of the most popular of all WRC models to date and was originally developed in Delphi. The success of the model was due in part to the simplicity of use and it remained unchanged for almost 10 years before being upgraded in 2008 (Figure 8.6). The new version contains virtually no new features and is effectively a face-lift which allows the model to operate under the various new operating systems which created certain conflicts with colour schemes and so on. The revised version should operate properly on any new Microsoft operating system and is available freely from the WRC web site (www.wrc.org.za).

The analysis of Minimum Night flow is discussed in more detail in Section 4 and a full discussion on how to interpret the results is provided through a series of annotated examples of real logging results.

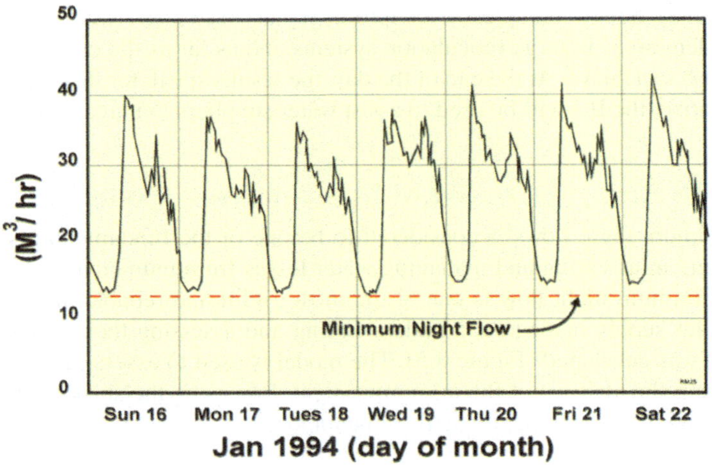

Figure 8.6 Example of inflow to a zone showing the minimum night flow.

Measurement of minimum night-flow into a zone-metered area is possibly one of the simplest and most valuable actions that a water supplier can take in order to identify whether or not they have a serious leakage problem.

The model is based directly on the BABE principals as set out by the UK Water Industry (Lambert, 1994) and is written in DELPHI for the Windows operating system. The SANFLOW model has been designed in such a way that it is extremely simple to use and all of the detailed calculations are hidden from the main screen. The main screen therefore provides a clear and concise overview of the leakage in a particular zone. Details for any of the calculated values can be viewed by simply selecting the variable from the main task bar.

A useful feature included in SANFLOW was the incorporation of a sensitivity analysis which can be used to test the sensitivity of the results to the reliability associated with certain parameters used in the calculations. The issue of the sensitivity of the various parameters and reliability of certain data have now become common practice in most BABE models. There was always a concern regarding the selection of certain process parameters or variables that could influence the overall leakage predictions. To establish if such variables had an important influence on the results, the user would normally have to change each variable individually and re-run the programme. In the case of SANFLOW, the sensitivity feature allows the user to view (graphically) the significance of changing the various process parameters either individually or simultaneously. This feature helps to address the common criticism of the basic BABE approach in which many users are concerned that the use of estimated parameters will jeopardize the results. It is often interesting to use the sensitivity feature to identify which, if any, parameters have a significant influence on the overall result.

To explain the use of SANFLOW, a typical normal inflow to a zone is shown in Figure 8.6 (UK Water Industry, Lambert, 1994), from which the minimum night flow can be identified as the lowest flow entering the zone at any time. In most zones the minimum night flow occurs sometime between midnight and 4 am. In order to evaluate the level of leakage in a particular zone from the inflow, as shown in Figure 8.6, the minimum night-flow is split into various components (some of which are pressure dependent) in accordance with the general BABE principles. Figure 8.7 shows the different components making up the minimum night-flow and these are fully explained in the SANFLOW user guide (WRC, 1999).

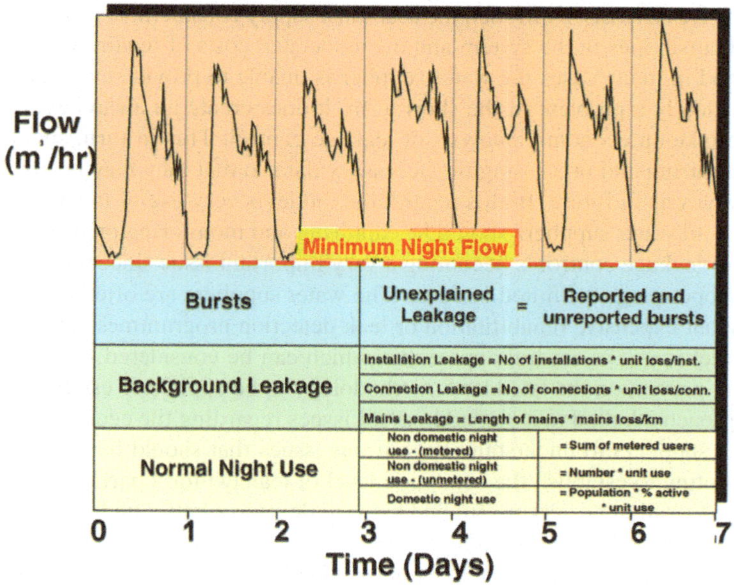

Figure 8.7 Components making up the minimum night flow.

The analysis of background night flows is comparatively simple and the SANFLOW model provides a quick and effective aid to water suppliers in this regard.

8.1.4 ECONOLEAK: Economics of leakage model

The ECONOLEAK model (Figure 8.8) is not designed to address the economic issues associated with all of the various types of leakage-reduction activities mentioned above. Instead, it is aimed specifically at determining when a water supplier should invest in active leakage control for a specific zone metered area. The model will assist water suppliers in gaining a better understanding of the main factors influencing the economics of leakage control and will enable them to identify the most cost-effective methods of reducing their system leakages.

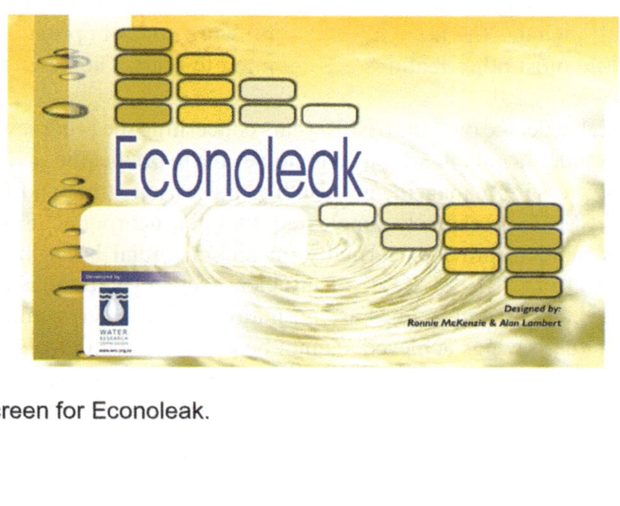

Figure 8.8 Opening screen for Econoleak.

In order to use the ECONOLEAK Model, the user must supply considerable factual system data relating to the frequency of burst pipes in the system and the associated costs of repair. Such information is often difficult to obtain and in many cases the water supplier is unable to provide even the most basic leakage data. While this is clearly a problem in the short term, it does create an awareness of what information is required to undertake an economic analysis of leakage control. This in turn, should encourage water suppliers to start capturing and processing the necessary data so that they may be able to carry out some form of economic analysis in future. In this regard, the model is very useful in creating awareness of the key information that all water suppliers should be capturing and monitoring on a continuous basis.

The economics of leakage control is becoming a very important issue since most water supply utilities in South Africa are operating on limited budgets. The water suppliers are often unable to provide proper motivation to carry out expensive rehabilitation or leak detection programmes. The approach used in the ECONOLEAK Model is one of several approaches which can be considered. It appears that there is no general consensus at present on the 'standard' methodology to be applied to establish the economic level of leakage. The approach adopted may not address all issues regarding the economics of leakage. It does, however, help water suppliers to understand the various issues that should be considered and taken into account when attempting to establish the economic level of leakage for a particular water supply system. In this regard, it has proved to be very useful and helps to develop a better understanding of the economic issues among water suppliers throughout South Africa.

The ECONOLEAK model enables a water supplier to identify when it is necessary to intervene through active leakage control. In other words, the programme will assist water suppliers in identifying when they should send a leak-detection and repair crew into an area to find unreported bursts. In order to use the model, the water supplier is required to gather the information indicated in Table 8.3.

It should be noted, that if the information is not readily available from the water supplier's records, the default values provided in the User Guide can be used until more reliable information is obtained (Mckenzie, 2002). The model uses the basic information described in Table 8.2 to provide the water supplier with an indication of when they should intervene in a particular zone and also how much funding should be allocated to leakage detection and repair per annum.

8.1.5 PRESMAC: Pressure management model

PRESMAC is a relatively simple tool used to provide an initial estimate of the likely savings that could be achieved in an area through some form of pressure management. The original model was developed in Delphi and allowed the user to analyse fixed outlet and time modulated pressure control. The latest version of PRESMAC (Figure 8.9) includes the flow modulated pressure control option, which was not available in the original model, while most other features are common to both. The new version is available from the WRC web site (www.wrc.org.za).

There has been considerable debate over the past 10 years concerning the accuracy of the predicted savings from any of the numerous models available which claim to be able to predict the savings that can be achieved through different forms of pressure management, including the various time and flow modulated options. The approach used in PRESMAC is relatively simple and does not incorporate a full hydraulic reticulation model, which many other vendors provide: normally on a commercial basis. In practice, however, it is usually found that the results from PRESMAC are of sufficient accuracy to establish whether or not pressure management is viable and the predicted savings are normally within the error margins expected. In most cases, it is found that the actual savings achieved, cannot be predicted with greater accuracy due to a wide range of factors that are excluded from any available model, such as quality of workmanship when laying the reticulation. The model does not include the savings that result from the reduced number of new burst pipes which can be highly significant and it also excludes the financial benefits resulting from the extended life-span of the reticulation system. This latter item can in many instances dominate the economics of the pressure management initiatives but it is also the most difficult item to quantify with any level of confidence. By excluding some of the benefits from the assessment, the results will err on the conservative side.

The PRESMAC pressure management model is used to assess the likely savings (in monitory terms) of various pressure reduction options (fixed outlet and time-modulated PRVs) in a selected zone metered area. The analysis is undertaken in a relatively simple and pragmatic manner based on the general BABE concepts. This approach allows the user of the programme to gauge the potential for pressure management very quickly and effectively without requiring a full detailed pipe network analysis. Although the methodology is based on a number of simplifications and assumptions, in practice, the predicted savings are generally within 10% to 20% of those actually achieved.

Table 8.3 Basic information required to run ECONOLEAK.

Description	Units	Default
Number of service connections	Number	–
Length of transmission mains	km	–
Length of distribution mains	km	–
Average system pressure	m	–
Unavoidable connection losses at 50 m of pressure	Litres/connection/hr	1.25
Unavoidable mains losses at 50 m of pressure	Litres/km/hr	20
Leakage from service reservoirs	% of volume per day	0.1
Leakage through mains burst	m^3/hr at 50 m pressure	12.0
Leakage from connection pipe burst	m^3/hr at 50 m pressure	1.6
Average running time of mains burst	Days	0.5
Average running time of connection pipe burst	Days	10
Average cost of repairing mains burst	Rand	3000
Average cost of repairing connection pipe burst	Rand	2000
Monthly water supplied to the zone or district	kilolitres	–
Estimated monthly real losses	kilolitres	–
Purchase price of water from bulk supplier	Rand/m^3	
Selling price of water	Rand/m^3	
Service connection bursts per 1000 connections at 50 m of pressure	Bursts/1000 conn/yr	2.5
Annual frequency of mains bursts per km of mains at 50 m of pressure	No./km of mains/yr	0.15
Pressure leakage exponent for flow through mains and connection leaks	–	0.7
Power exponent for calculating number of mains leaks for different pressures (cubic relationship is normally adopted)	–	3
Cost of basic sounding per km of mains	Rand/km mains	700
Cost of leak noise correlator per km of mains	Rand/km mains	1400
% of mains requiring leak noise correlator to detect leaks	%	20

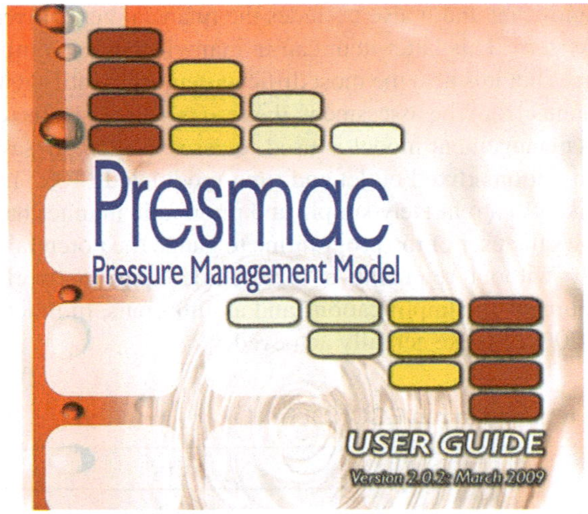

Figure 8.9 Opening screen from the PRESMAC pressure management model.

8.1.6 WDM scorecard

Before implementing any new WDM intervention, it is normal practice to undertake an investigation of the area in question in order to assess the key problems and to propose a strategy to address them. The resulting WDM strategy is usually a range of actions or interventions designed to address the main problems in order to reduce losses and or consumptive use.

Balanced scorecards have been used for many years to assess and monitor complex situations which involve a wide range of functions, many of which cannot be assessed or quantified in a normal manner. It is ideally suited to multi-disciplinary activities such as the operation and management of a water utility for example. In such an organization there may be various technical issues that must be assessed and monitored as well as different human resources activities and different management activities, and so on. all of which are important in their own right but cannot be directly compared with each other. It is therefore difficult to evaluate and measure the overall performance of the utility without resorting to some form of balanced scorecard approach. In 1994, a balanced scorecard methodology was adopted in order to evaluate the many aspects of a WDM strategy for a large water supplier. The balanced scorecard was ideally suited to the evaluation and monitoring of the WDM activities, which included items such as education and awareness, leak location, water auditing, and so on. The methodology was quickly modified into a simple spreadsheet (WRC, 2010) comprising between 20 and 25 items, each of which carried a weighting of either 10 or 20 depending upon the importance of the specific item. The balanced scorecard approach was used to evaluate the WDM situation in many municipalities throughout South Africa and also several other countries around the world. It was found to be very helpful in developing very quick first order WDM strategies for all areas in which it was used. The respective clients also found the process extremely helpful since they were involved directly with the evaluation and scoring of each item.

The methodology is far from complicated and is very simple and straightforward to use. It is extremely flexible and can be modified to suit a specific application or client or even be colour coded if this is considered easier to understand. Many variations of the methodology have already been applied and the items included in the scorecard can be reduced or increased where appropriate. From the analysis of many utilities it was found that 20 to 40 items is generally more than sufficient to capture the key elements needed by a specific water supplier to develop a practical WDM strategy (Figures 8.10–8.13). When more items are included, it tends to become a more academic exercise with a risk of losing focus on the key problem areas.

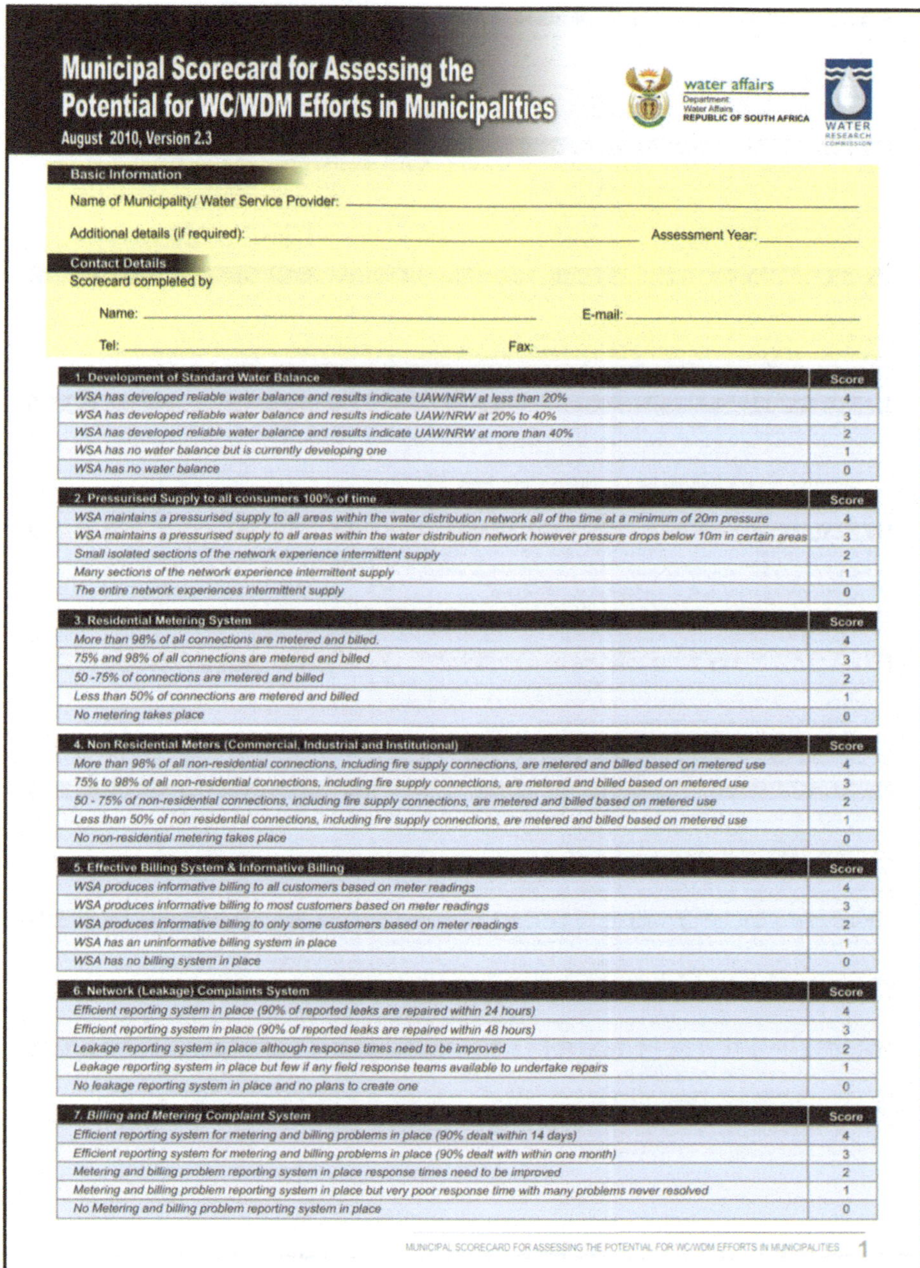

Municipal Scorecard for Assessing the Potential for WC/WDM Efforts in Municipalities

August 2010, Version 2.3

water affairs
Department: Water Affairs
REPUBLIC OF SOUTH AFRICA

WATER RESEARCH COMMISSION

Basic Information

Name of Municipality/ Water Service Provider: __________

Additional details (if required): __________ Assessment Year: __________

Contact Details

Scorecard completed by

Name: __________ E-mail: __________

Tel: __________ Fax: __________

1. Development of Standard Water Balance	Score
WSA has developed reliable water balance and results indicate UAW/NRW at less than 20%	4
WSA has developed reliable water balance and results indicate UAW/NRW at 20% to 40%	3
WSA has developed reliable water balance and results indicate UAW/NRW at more than 40%	2
WSA has no water balance but is currently developing one	1
WSA has no water balance	0

2. Pressurised Supply to all consumers 100% of time	Score
WSA maintains a pressurised supply to all areas within the water distribution network all of the time at a minimum of 20m pressure	4
WSA maintains a pressurised supply to all areas within the water distribution network however pressure drops below 10m in certain areas	3
Small isolated sections of the network experience intermittent supply	2
Many sections of the network experience intermittent supply	1
The entire network experiences intermittent supply	0

3. Residential Metering System	Score
More than 98% of all connections are metered and billed.	4
75% and 98% of all connections are metered and billed	3
50 -75% of connections are metered and billed	2
Less than 50% of connections are metered and billed	1
No metering takes place	0

4. Non Residential Meters (Commercial, Industrial and Institutional)	Score
More than 98% of all non-residential connections, including fire supply connections, are metered and billed based on metered use	4
75% to 98% of all non-residential connections, including fire supply connections, are metered and billed based on metered use	3
50 - 75% of non-residential connections, including fire supply connections, are metered and billed based on metered use	2
Less than 50% of non residential connections, including fire supply connections, are metered and billed based on metered use	1
No non-residential metering takes place	0

5. Effective Billing System & Informative Billing	Score
WSA produces informative billing to all customers based on meter readings	4
WSA produces informative billing to most customers based on meter readings	3
WSA produces informative billing to only some customers based on meter readings	2
WSA has an uninformative billing system in place	1
WSA has no billing system in place	0

6. Network (Leakage) Complaints System	Score
Efficient reporting system in place (90% of reported leaks are repaired within 24 hours)	4
Efficient reporting system in place (90% of reported leaks are repaired within 48 hours)	3
Leakage reporting system in place although response times need to be improved	2
Leakage reporting system in place but few if any field response teams available to undertake repairs	1
No leakage reporting system in place and no plans to create one	0

7. Billing and Metering Complaint System	Score
Efficient reporting system for metering and billing problems in place (90% dealt within 14 days)	4
Efficient reporting system for metering and billing problems in place (90% dealt with within one month)	3
Metering and billing problem reporting system in place response times need to be improved	2
Metering and billing problem reporting system in place but very poor response time with many problems never resolved	1
No Metering and billing problem reporting system in place	0

MUNICIPAL SCORECARD FOR ASSESSING THE POTENTIAL FOR WC/WDM EFFORTS IN MUNICIPALITIES 1

Figure 8.10 Page 1 from the WDM scorecard.

8. Asset Register for water Reticulation System	Score
WSA has a comprehensive and accurate asset register in place which is available digitally	4
WSA has a partially completed accurate asset register in place	3
WSA has a poor asset register in place	2
WSA is in the process of developing an asset register	1
WSA has no asset register in place and no immediate intention of generating one	0

9. Asset Managemeknt - Capital Works	Score
2% or more of the value of the water network is invested annually into new capital works related to the existing infrastructure	4
1% - 2% of the value of the water network is invested annually into new capital works related to the existing infrastructure	3
Less than 1% of the value of the water network is invested annually into new capital works related to the existing infrastructure	2
No estimate of asset value of water supply system is available but WSA feels that sufficient budget is spent on new Capital Works	1
No estimate of asset value of water supply system is available and WSA feels that insufficient budget is spent on new Capital Works	0

10. Asset Management - Operations and Maintenance	Score
2% or more of the value of the water network is invested annually into the maintenance of the existing infrastructure	4
1% - 2% of the value of the water network is invested annually into the maintenance of the existing infrastructure	3
Less than 1% of the value of the water network is invested annually into the maintenance of the existing infrastructure	2
No estimate of asset value of water supply system is available but WSA feels that sufficient budget is spent on operations and maintenance	1
No estimate of asset value of water supply system is available but WSA feels that insufficient budget is spent on operations and maintenance	0

11. Dedicated WDM Support	Score
Efficient WDM Section in place with sufficient resources	4
WDM section in place requires some resources and capacity building	3
WDM section in place. Major resources and capacity building required	2
No WDM section currently in place, although it is in process of being created	1
No WDM Section and no intention to create WDM section	0

12. Active Leakage Control	Score
Active leakage detection and repair undertaken continuously with average sweep time of 12 months or less	4
Active leakage detection and repair undertaken continuously with average sweep time of 48 months or less	3
Active leakage detection and repair is undertaken on an add-hoc basis	2
No active leakage is currently undertaken, however, the WSA intends to initiate such measures	1
No active leakage detection is undertaken and the WSA has no intention to conduct such measures	0

13. Effective Sectorisation	Score
Reticulation network has been sectorised and is checked regularly to maintain discrete zones	4
Reticulation network has been sectorised but is not checked regularly to ensure discrete zones	3
Only portions of the reticulation network have been sectorised	2
Few if any zones have been created but plans are in place to sectorise the system	1
No sectorisation has been undertaken and no plans are in place to implement such measures	0

14. Effective Bulk Meter Management	Score
All bulk water sources to the WSA are metered by the WSA using some form of check metering(either permanent or temporary)	4
All bulk water sources to the WSA are metered by the Bulk water supplier or by the WSA	3
Few bulk water meters are operational	2
No Bulk Metering in place, however, WSA has plans to install bulk meters	1
No Bulk metering in place and no plans for such meters have been made	0

15. Effective Zone Meter Management and Assessment of minimum	Score
All inlet points to discrete zones are metered and accurate with Minimum Night Flows logged and analysed on a regular basis	4
All inlet points to discrete zones are metered and accurate but no Minimum Night Flow analyses are undertaken	3
All inlet points to discrete zones are metered but many are broken or considered to be inaccurate	2
Zone inputs are currently not metered although the WSA has planned to install meters on all zone inlets	1
No accurate zone metering is in place and there are no plans to introduce such measures	0

16. Pressure Management and Maintenance of Pressure Reducing Valves	Score
Reticulation is comprehensively sectorised into pressure zones which are all discrete. All PRV's are maintained under maintenance schedule	4
Reticulation is comprehensively sectorised into pressure zones which are all discrete. PRV's are only maintained when problems become apparent	3
Reticulation is sectorised in pressure zones but the zones are not verified and little or no maintenance is undertaken on the PRV's	2
WSA intend to introduce pressure zones and the use of PRV's to manage system pressures	1
No Discrete pressure zones and no PRV maintenance	0

MUNICIPAL SCORECARD FOR ASSESSING THE POTENTIAL FOR WC/WDM EFFORTS IN MUNICIPALITIES 2

Figure 8.11 Page 2 of the WDM scorecard.

17. As-built Drawings of Bulk and Reticulation Infrastructure	Score
Accurate as-built drawings for all reticulation are available digitally	4
As-built drawings available digitally for the majority of the network and available in hard copy for the remainder of the network	3
A mixture of digital and hard copy as-built drawings available for the majority of the network but many problems are known to exist with the data quality	2
Only some hard copy as-built drawings are available for portions of the network	1
No as-built drawings available	0

18. Schematic Layout of Water Infrastructure	Score
An up-to date and detailed schematic of the whole bulk reticulation network is available	4
A detailed schematic of the bulk reticulation network is available but is known to be outdated and/or inaccurate	3
Only a rough schematic of the bulk reticulation network is available which is known to be inaccurate and/or outdated	2
No schematic of the bulk reticulation is available although the WSA is planning to develop such a schematic	1
No schematic of the bulk reticulation is available and the WSA has no plans to develop such a schematic	0

19. Regulations and By-laws	Score
Regulations and By- laws are in place which address WDM issues and some form of enforcement is undertaken	4
Regulations and By-laws are in place which address WDM issues but are not enforced	3
Regulations and By-laws are in place but do not address WDM issues	2
There are no By-laws in place but SA is intending to introduce such measures	1
There are no By-laws in place and WSA has no plans to introduce them	0

20. Tariffs	Score
WSA has rising block tariffs in place that encourage water use efficiency	4
WSA has rising block tariffs in place but they do not encourage water use efficiency sufficiently	3
WSA has single water tariff in place	2
WSA has a declining block tariff in place	1
WSA does not know what tariff structure is in place	0

21. Technical Support to Customers	Score
The WSA actively engages with customers and offers technical support on WDM to both domestic as well as commercial/industrial customers	4
The WSA offers technical support on WDM to large consumers on a pro-active basis	3
The WSA only offers technical support on WDM on a reactive basis	2
The WSA currently offers no technical support but plans to introduce a support mechanism	1
The WSA has no plans to offer technical support on WDM measures to any customer	0

22. Removal of Unlawful Connections	Score
The WSA actively monitors and removes all unlawful connections	4
The WSA selectively monitors and removes unlawful connections	3
The WSA monitors unlawful connections but has no policy for removal	2
The WSA plans to introduce measures to tackle unlawful connections	1
The WSA has no plans to deal with unlawful connections	0

23. Community Awareness and Education Programmes	Score
WSA is actively involved in conducting workshops on water conservation within the communities with a dedicated team	4
WSA is involved in conducting workshops on water conservation within the communities however no dedicated team exists	3
WSA has very little involvement with workshops on water conservation within the communities	2
WSA currently does not conduct workshops on water conservation within the communities, however these interventions are proposed	1
WSA currently does not conduct workshops on water conservation within the communities	0

24. Schools Awareness and Education Programmes	Score
WSA is actively involved in conducting workshops on water conservation within the schools with a dedicated team	4
WSA is involved in conducting workshops on water conservation within the schools however no dedicated team exists	3
WSA has very little involvement with workshops on water conservation within the schools	2
WSA currently does not conduct workshops on water conservation within the schools, however these interventions are proposed	1
WSA currently does not conduct workshops on water conservation within the schools	0

25. Newspaper & radio articles plus posters and leaflets for distribution	Score
WSA runs regular adds in newspapers and/or radio and has library of posters and leaflets for public distribution	4
WSA runs occasional adds in newspapers and/or radio and has library of posters and leaflets for public distribution	3
WSA has library of posters and leaflets for public distribution but does not advertise in newspapers or radio	2
WSA has some leaflets and/or posters and intends to strengthen its capacity to promote WC/WDM in the community	1
WSA does not advertise in newspapers or radio and has no posters or leaflets on WC/WDM	0

MUNICIPAL SCORECARD FOR ASSESSING THE POTENTIAL FOR WC/WDM EFFORTS IN MUNICIPALITIES 3

Figure 8.12 Page 3 of the WDM scorecard.

"The purpose of the Municipal Scorecard for the assessment of Water supply systems is to ascertain the Status quo of these systems and evaluate the potential for Water Conservation / Water Demand Management (WC/WDM) measures to be implemented in these systems. The scorecard is designed to enable the Regulator (Department of Water Affairs) to assess the current situation regarding losses and levels of wastage in all water supply systems countrywide. The initiative will provide the mechanism whereby the DWA can identify areas where WC/WDM is not being addressed properly to ensure that appropriate measures are taken to encourage efficient use of water and the elimination of wastage. The scorecard is based on a number of key issues which all water service providers must address as part of normal management. The Scorecard will also identify areas of expertise and best practice which can be used to help other areas which are experiencing problems. The following table provides the items included in the Scorecard."

Name of Municipality/ Water Service Provider: ______________________

Item No.	Description	Points
1.	*Development of Standard Water Balance*	
2.	*Pressurised supply to all consumers 100% of time*	
3.	*Residential Metering System*	
4.	*Non Residential Meters (Commercial, Industrial and Institutional)*	
5.	*Effective Billing System & Informative Billing*	
6.	*Network (Leakage) Complaints System*	
7.	*Billing and Metering Complaints System*	
8.	*Asset Register of Water Reticulation System*	
9.	*Asset Management - Capital Works*	
10.	*Asset Management - Operations and Maintenance*	
11.	*Dedicated WDM support*	
12.	*Active Leakage Control*	
13.	*Effective Sectorisation*	
14.	*Effective Bulk Meter Management*	
15.	*Effective Zone Meter Management and Assessment of Minimum Night Flows*	
16.	*Pressure Management and Maintenance of Pressure Reducing Valves*	
17.	*As-Built Drawings of Bulk and Reticulation Infrastructure*	
18.	*Schematic Layout of Water Infrastructure*	
19.	*Regulations and By-Laws*	
20.	*Tariffs*	
21.	*Technical Support to Customers*	
22.	*Removal of Unlawful Connections*	
23.	*Community Awareness and Education Programmes*	
24.	*Schools Awareness and Education Programmes*	
25.	*Newspaper & radio articles plus posters and leaflets for distribution*	

Total:

4 MUNICIPAL SCORECARD FOR ASSESSING THE POTENTIAL FOR WC/WDM EFFORTS IN MUNICIPALITIES

Figure 8.13 Summary page for the WDM scorecard.

8.2 LATEST RESULTS FROM THE NRW ASSESSMENTS

Using the various models developed by the WRC, a series of NRW assessments has been undertaken to establish the likely level of NRW throughout the whole country. The first NRW assessment was undertaken in 1999 and was based on only 20 datasets which were considered to be of an acceptable quality from a potential set of approximately 600 water suppliers. The assessment suggested that the average NRW for the 20 water suppliers was in the order of 25% with an average ILI value of 6.0. Most of the acceptable datasets were provided from the larger municipalities, which were the only water suppliers at this time who collected the appropriate base data and meter readings. As a result of this initial assessment, the WRC commissioned a follow-up assessment in 2005.

In the 2005 assessment (WRC, 2005) information from 60 water suppliers was obtained from which 30 acceptable datasets were identified representing just under 50% of the total municipal water supplied throughout South Africa. In this assessment, the percentage NRW was not calculated in line with the IWA recommendations on avoiding the use of percentages when dealing with NRW. The ILI which provides an indication of the physical leakage was however calculated for the 30 municipalities and an average value of 6.3 was derived. Once again, the value of the assessment was clear to the Water Research Commission as well as the Government which commissioned a third assessment to be undertaken.

The third assessment was undertaken in 2007 (WRC, 2007) and involved 100 datasets from which 62 were included in the final assessment representing almost 60% of the total municipal water use in South Africa. In this assessment, many of the smaller municipalities were included and the NRW was estimated to be 36% with an average ILI of 7.6. The percentage NRW was again included in the assessment despite the fact that it was accepted that percentages can be very misleading. Some of the high level committees and politicians were uncomfortable with the use of the ILI and the various other recommended performance indicators. As a result, they insisted on the use of percentages albeit with a 'health warning' to highlight that they can be misleading in certain cases.

Following the success of the 2007 assessment in raising the issue of Non-Revenue Water to a national platform where it was discussed at length by Government, a fourth assessment was undertaken between 2010 and 2012, the results of which were officially released in March of 2013 (WRC, 2012). This assessment is the most comprehensive and detailed assessment of NRW undertaken in South Africa and involved water balance information from more than 130 municipalities. The project was supported not only by the Water Research Commission but also the Department of Water Affairs.

The data gathered from 132 of the possible 237 municipalities supplying water to more than 40 million residents throughout South Africa represents over 75% of the total volume of municipal water supply. The results indicate that the current level of Non-Revenue Water estimated for the country as a whole is almost 37% (as shown in Figure 8.14) with an average ILI of 6.8.

It should be noted that although the general trend in NRW appears to be increasing, as can be seen in Figure 8.14, the results are influenced by a number of factors which must be carefully considered before reaching such a conclusion. The most important consideration is the size and reliability of the dataset used in the analyses. The size and reliability of the dataset has been increasing with each new assessment and it is clear that the additional datasets that are being added annually tend to be the more problematic municipalities that experience much higher levels of leakage than those which have been assessing their NRW levels for the past 10 years or more. With each new assessment, the datasets are being assessed in more detail and various errors and problems with interpretation of the IWA water balance are being corrected. Invariably this tends to push more water into the NRW component, which would explain some of the increase shown in Figure 8.14. As all of the metros and most of the larger cities in South Africa have now been included in the assessment, the database should now be stable and the true trend in NRW will be confirmed over the next few assessments. Establishing the status quo has been a significant challenge for a country with a population of over 50 million and the most recent assessment has created a solid base from which the future assessments and trends can be determined.

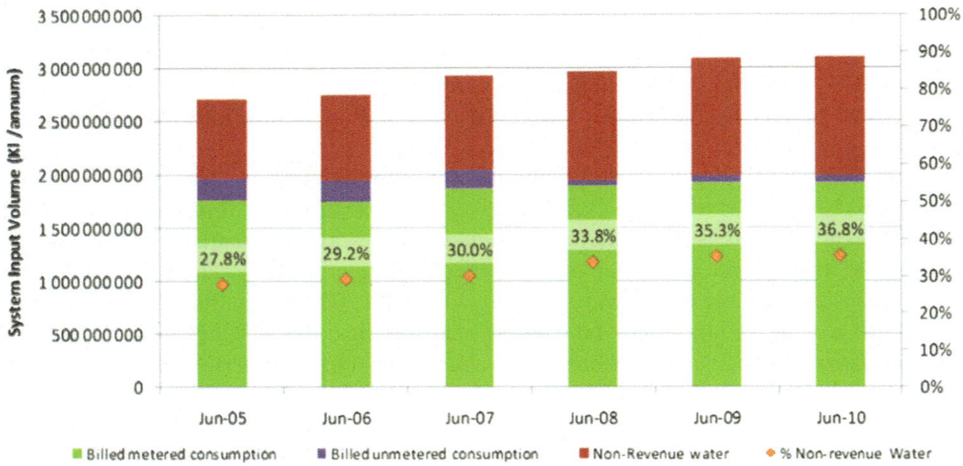

Figure 8.14 Growth in NRW in South Africa.

Although the NRW figure for South Africa of 37% is high, as it suggests that more than one third of all water supplied to municipalities is not billed to consumers, this figure is broadly in line with the world average of ±36.6%, as can be seen in Figure 8.15.

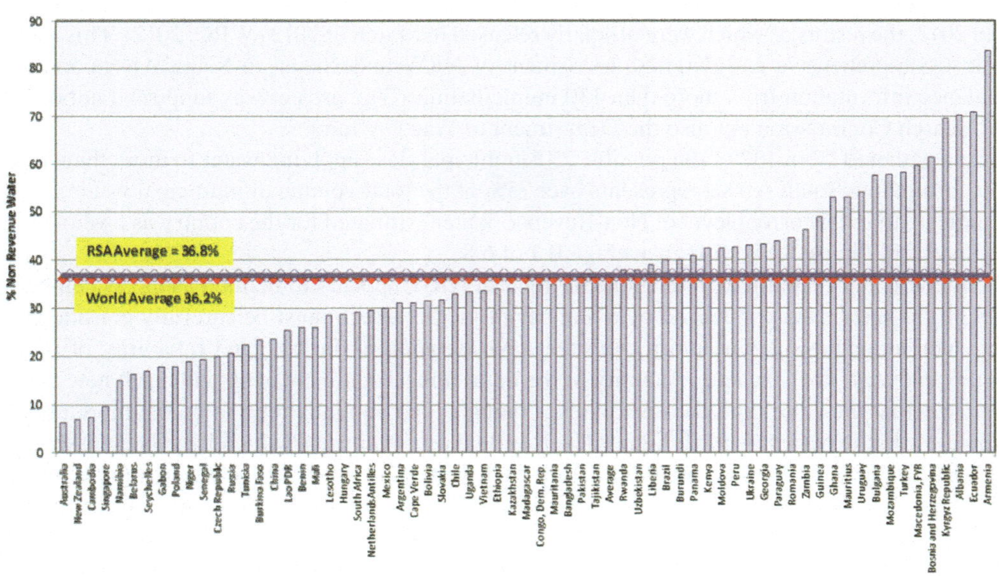

Figure 8.15 South African NRW compared to world datasets.

As can be seen from the figure, the NRW for South Africa appears high if compared to other developed countries but appears low if compared to most developing countries. In this regard South Africa tends to be an anomaly as it has both developed and developing areas which in turn can experience NRW levels as low as some of the best in the world, while in nearby developing areas the NRW levels can be up at ±90%. The figures are further complicated by the fact that there may be dense urban areas adjacent to rural areas with scattered communities and to compare the NRW levels using percentages in such cases is meaningless.

To try and overcome the problems of using percentages in South Africa, the IWA recommended performance indicator for physical leakage called the ILI is used. South Africa was one of the first countries outside of the UK (where the ILI index was developed – see Lambert *et al.*, 1999) to embrace the use of the ILI as the key leakage indicator. The ILI of 6.8 for the country as a whole, is considered to provide a realistic indicator of physical leakage for the South African systems. It is interesting to note that the various estimates of ILI over the past 12 years have all been between 6 and 8. Again, such leakage would be considered high for most developed countries but low for most developing countries. The ILI highlights the fact that levels of physical leakage are relatively high in South Africa and there is considerable scope for improvement. This is both positive and negative for the country in that the leakage levels are high, which has a negative connotation, however, such high leakage represents a 'resource' which can be accessed through proper management and better water use practices. This aspect is highlighted in Figure 8.16 which provides an indication of the per capita consumption in South Africa.

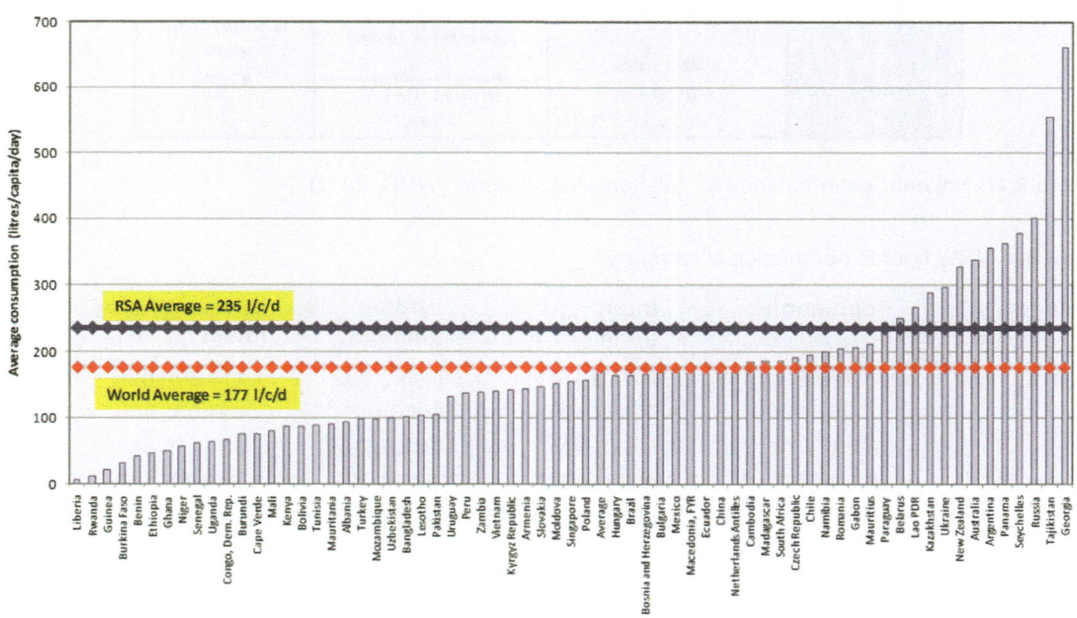

Figure 8.16 Estimated per capita water consumption in South Africa.

Although the estimates provided in Figure 8.15 must be considered as relatively course estimates, they highlight that the per capita water consumption in South Africa is significantly higher than the world average despite the fact that South Africa is a water scarce country and should exhibit a per capita consumption below the world average. Once again the figures tend to support the general conclusion that water is being wasted in the country and there is significant scope for improvement. In Australia for example, during the recent drought, the residents in many water stressed parts of the country reduced their average per capita water use to less than 140 l per day. It can be achieved, and the scope for reducing water use in South Africa is significant and provides some level of assurance to support future growth in certain areas where new water resources cannot be developed or are simply unavailable.

The most recent water balance developed for the municipal water use in South Africa is shown in Figure 8.17, which is based on the standard International Water Association guidelines. It should be noted that in South Africa, every water supplier is categorized according to the size of the population supplied and whether the area is urban or rural. The results from the breakdown into the different categories are provided in Table 8.4 from the 2012 assessment and the category definitions used in the assessment for all water suppliers countrywide are given in Table 8.5.

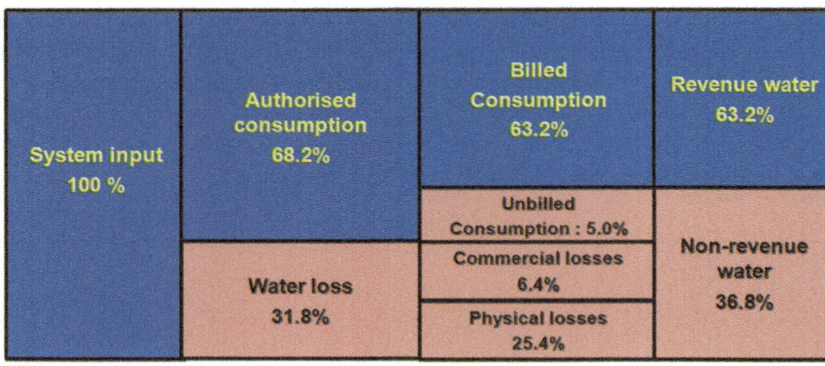

Figure 8.17 National water balance for SA from WRC Report (WRC, 2012).

Table 8.4 NRW figures per municipal category.

Category	Population	Input (m^3/a)	NRW (m^3/a)	Revenue water (m^3/a)	l/c/d
A	17,420,512	1,849,091,117	634,192,022	1,214,899,095	291
B1	7,756,187	683,667,320	282,585,164	401,082,156	241
B2	3,882,070	325,623,095	99,407,207	226,215,889	230
Urban Total	**29,058,769**	**2,858,381,532**	**1,016,184,393**	**1,842,197,140**	**269**
B3	3,845,279	230,642,568	85,229,869	145,412,699	164
B4	4,245,736	101,138,956	73,334,514	27,804,442	65
Rural Total	**8,091,015**	**331,781,524**	**158,564,383**	**173,217,141**	**112**
National Total	**37,149,784**	**3,190,163,056**	**1,174,784,776**	**2,015,414,281**	**235**
Extrapolated	**49,988,373**	**4,292,650,981**	**1,580,730,012**	**2,711,920,969**	**235**

In view of the fact that the 2012 assessment was unable to obtain data from every water supplier in the country, it was still necessary to extrapolate the results in order to derive an estimate of NRW for the whole country. Fortunately, the available datasets gathered in the 2012 assessment covered ± 37 million of the estimated ±50 million South Africans and therefore the extrapolation was used to represent the missing ±13 million. The results from this are shown in the last row of Table 8.4.

Table 8.5 Definitions of municipalities.

Category	Number	Short description	Long description
A	8	Metros	Metropolitan municipalities
B1	19	Major cities	Secondary cities, local municipalities with the largest budgets
B2	27	Minor cities	Municipalities with a large town as core
B3	110	Rural dense	Municipalities with relatively small population and significant proportion of urban population but with no large town as core
B4	70	Rural scattered	Municipalities which are mainly rural with, at most, one or two small towns in their area
C1			District municipalities which are not water services providers
C2			District municipalities which are water services providers

From the latest NRW study undertaken by the Water Research Commission in partnership with the Department of Water Affairs, it was possible to gather information for more than 75% of the water supplied throughout South Africa. The overall NRW for South Africa is estimated to be ±1580 million m^3/annum, which is approximately one third of the total water supplied. Conservatively, this represents a loss of over ±R7 billion (almost $1 billion) based on an average bulk water tariff of approximately R5/m^3.

The average ILI value for all of the South African Municipalities was estimated to be 6.8, which again is in line with the world average and would be above average (i.e., bad) when compared to most developed countries and well below average (i.e., good) when compared to most developing countries. Effectively, the ILI value of 6.8 tends to support the perception created from the percentage non-revenue water figures for South Africa (36.8%) where there is clearly a high level of wastage or water losses in the country and considerable scope for improvement.

The above figures are based on the Standard IWA Water Balance in which the 'Revenue Water' figures provided by the financial departments are assumed to be correct. In South Africa, however, there can be a significant component of revenue water which is billed but never paid for by the consumers. Preliminary estimates of this component suggest that if it is taken into account, the level of NRW may increase by up to 10%. Investigations are continuing to try and quantify this element with greater reliability so that the next assessment can provide a more complete and accurate water balance.

BIBLIOGRAPHY

Bhagwan J. (2014). Counting the Lost Drops – SA's Study into Non-Revenue Water. Paper presented at the *Singapore International Water Week*, 1–5 June.

Bhagwan J. and Mckenzie R. (2009). Water Research Commission's Role in Supporting WDM Interventions in South Africa. Paper presented at the *IWA Water Losses Specialist Group*, Bi-annual Conference in Cape Town, 26–30 April.

Lambert *et al.* (1994). *The 'Managing Leakage' Series of Reports*

- Report A : Summary Report. ISBN: 1 898920 06 0
- Report B : Reporting Comparative Leakage Performance. ISBN: 1 898920 07 9
- Report C : Setting Economic Leakage Targets. ISBN: 1 898920 08 7
- Report D : Estimating Unmeasured Water Delivered. ISBN: 1 898920 09 5
- Report E : Interpreting Measured Night Flows. ISBN: 1 898920 10 9
- Report F : Using Night Flow Data. ISBN: 1 898920 11 7
- Report G: Managing Water Pressure. ISBN: 1 898920 12 5
- Report H: Dealing with Customers' Leakage. ISBN: 1 898920 13 3
- Report J: Leakage Management Techniques, Technology and Training. ISBN: 1 898920 14 1

Lambert A. O. (1997). Pressure Management/Leakage Relationships: Theory, Concepts And Practical Applications. *Conference on Minimising Losses in Water Supply Systems*, April, IQPC Ltd, London.

Lambert A., Brown T. G., Takizawa M. and Weimer D. (1999) A review of performance indicators for real losses from water supply systems. *Aqua*, **48**(6), ISSN 0003-7214.

May J. (1994) Pressure dependent leakage, *World Water and Environmental Engineering*, October.

Mckenzie R., Bhagwan J. and Lambert A. (2002). Leakage Reduction Software Developed through the Water Research Commission. Paper presented at the *IWA Water Losses Task Force Managing Leakage Conference*, Cyprus, November.

SABS (1999). *South African Code of Practice for the Management of Potable Water in Distribution Systems*. SABS: 0306, ISBN 0-626-12046-2.

WRC (1999). *Development of a Standardised Approach to Evaluate Bursts and Background Losses in Water Distribution Systems in South Africa: SANFLOW.* Report TT109/99, by R. Mckenzie to the South African Water Research Commission, June, ISBN No. 1 86845 490 8.

WRC (2001). *Development of a Pragmatic Approach to Evaluate the Potential Savings from Pressure Management in Potable was Distributions in South Africa: PRESMAC.* Report TT152/01 by R. Mckenzie published by the South African Water Research Commission, July, ISBN No. 1 86845 722 2.

WRC (2002a). *Development of a Windows Based Package for Assessing Appropriate Levels of Active Leakage Control in Potable Water Distribution Systems: ECONOLEAK.* Report TT 169/02 published by the South African Water Research Commission, April, ISBN No. 1 86845 832 6.

WRC (2002b). *Development of a Simple and Pragmatic Approach to Benchmark Real Losses in Potable was Distribution Systems in South Africa: BENCHLEAK.* Report TT159/01 by Mckenzie & Lambert published by the South African Water Research Commission, January, ISBN No. 1 86845 773 7.

WRC (2005). *Benchmarking of Leakage from Water Reticulation Systems in South Africa.* Report prepared WRC by Mckenzie R. S. and Seago C. J. Water Research Commission Report Number TT 244/05, ISBN 1-77005-282-8.

WRC (2007). *An Assessment of Non-Revenue Water in South Africa.* Report prepared for DWAF and the WRC by C. J. Seago and R. S. Mckenzie Water Research Commission Report Number TT 300/07, ISBN 978-1-77005-529–2.

WRC (2009). *Aqualite Water Balance Software: User Guide.* Report prepared for the WRC by R. Mckenzie. Water Research Commission Report Number TT 315/07, ISBN 978-1-77005-599-5, August.

WRC (2010). *Municipal WDM Balanced Scorecard Model: User Guide.* Report prepared for the WRC by R. Mckenzie. Water Research Commission Report Number TT 523/12, ISBN 978-1-4312-0271-3, September.

WRC (2012). *The State of Non-Revenue Water in South Africa.* Report prepared for DWAF and the WRC by R. S. Mckenzie, Z. Siqalaba and W. Wegelin Water Research Commission Report Number TT 522/12, ISBN 978-1-4312-0263-8.

Chapter 9

International Case Studies from around the World

Chapter 9.1

Application in water loss reduction project in Bosnia and Herzegovina

Đevad Koldžo and Branko Vučijak

9.1.1 INTRODUCTION

Tuzla is north-eastern city in Bosnia and Herzegovina famous for salt production since olden times. But salt exploitation permanently causes soil subsidence, which causes leaks as a consequence. That was the basis for KfW bank initiating its water loss reduction project in one selected DMA, within the frame of the Master Plan project for Tuzla municipality. DMA, Kuzici having almost 1100 connections and more than 13 km of water supply network.

During the project, two cycles of the hydraulic measurements were performed so that the relevant conditions for the calculation of water balance in-line with the IWA methodology using a 'bottom-up' approach could be used.

Although the top-down approach is much cheaper and faster (the reasons it is mostly preferred in Bosnia and Herzegovina), and this approach is still based on a large number of estimates provided by the water utility companies, results obtained are usually unacceptable and incorrect. Therefore, for the majority of projects on water loss reduction, water loss is calculated on the basis of data obtained by hydraulic measurements of flow and pressure, that is, by using the bottom-up approach.

A major problem in applying the 'bottom-up' approach related to lack of adequate software to be used for all projects that would speed up the water balance evaluations. Recently, HEIS was involved simultaneously in nine projects on the water loss reduction during the same period. Taking that into account, the necessity to create such software was stressed and the decision was made to develop such software. Figure 9.1.1 presents CalcuLEAKator software developed in Bosnia and Herzegovina.

9.1.2 WATER BALANCE EVALUATION SOFTWARE – CALCULEAKATOR

The CalcuLEAKator Program was created in order to enable an easy and fast calculation of the water balance elements and success performance indicators by the 'bottom-up' approach. The tool was made using the widely known MS Excel program environment, which enables a wide range of users to use it. CalcuLEAKator was developed as a multilingual tool, presently enabling Bosnian and English language environment selection, while it is planned in the future to be enriched by an option of adding a language according to the needs of the user.

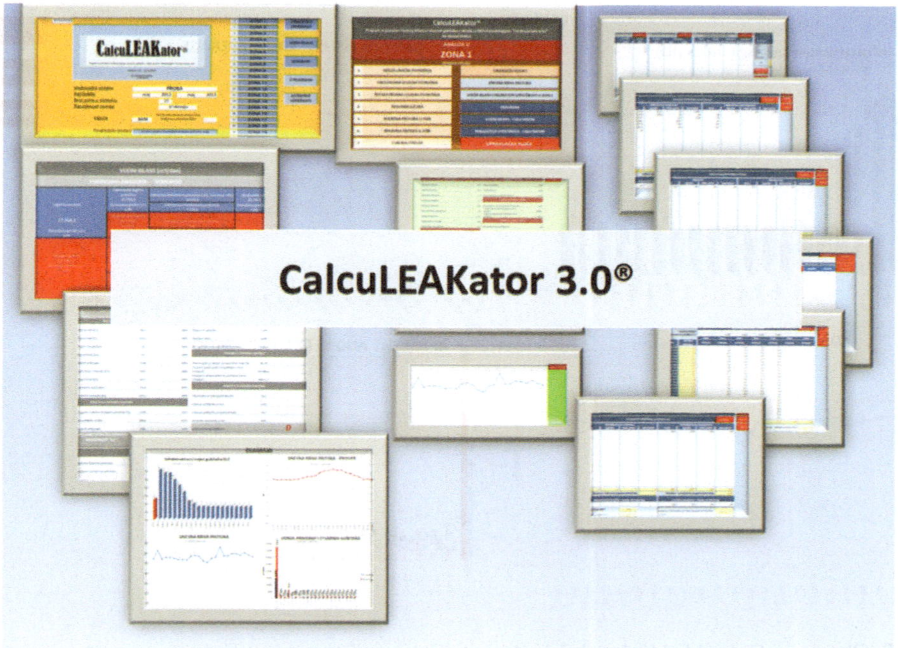

Figure 9.1.1 CalcuLEAKator 3.0 software (*Source*: Đevad Koldžo).

CalcuLEAKator will, except in the analytical form, present the desired results as graphs as well, as an example, it is possible to see a daily flow curve for each DMA separately, as well as a daily flow curve for the whole system. The program will present in the form of a diagram the values of a key technical indicator of success: Infrastructure Leakage Index (ILI) for the whole system, as well as for each of the DMAs sorted (see Figure 9.1.2).

The ILI represents the ratio between measured annual real losses and unavoidable annual current losses. This index is a key index for 'benchmarking', and based on ILI and according to the recommendations of the IWA CalcuLEAKator will define a belonging success category (four categories, from A (the best) to D (the worst)), and present the general conclusion with recommendations for further activities. Except as already mentioned, the program will present in the form of a diagram average hourly values for the pressure within the system, as well as the ratio between the values of real and apparent losses in each of the measured DMAs within the whole system.

The accuracy of the data depends on the accuracy of the data entered into the program. Regarding the fact that the 'bottom-up' approach is based on the measured results, the accuracy of the results will therefore as far as possible depend on the accuracy of the measured values.

Allowed error in the program CalcuLEAKator is defined trough the 95% safety of the accuracy, established as a procedure through which the evaluation of the level of uncertainty of particular components of the water balance is done. This 95% margin of the safety originally comes from the uncertainty of the calculation, and it is based on the normal distributions characteristics whereby 95% of the measurement belongs to the range of ±1,96 of the standard deviations (σ) around the mean value.

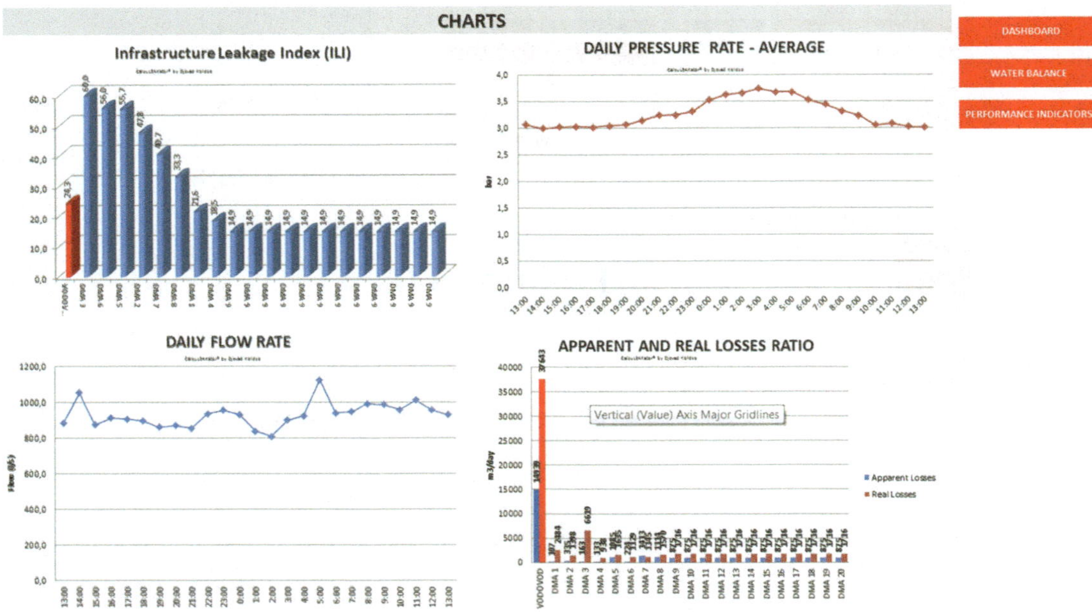

Figure 9.1.2 Charts in CalcuLEAKator 3.0 software (*Source*: Đevad Koldžo).

9.1.3 WATER LOSS REDUCTION PROJECT IN DMA KUZICI (TUZLA)

At the beginning of the project, the DMA was visited in order to inspect the possibilities of its isolation, and to find adequate points for the installation of a portable ultrasonic flow meter and pressure probes for measuring the DMA inflow and water pressure within the system.

At the same time, a GPS was used to record the positions of all border points in the DMA in order to verify the altitudes, and it was determined that the DMA is situated at altitudes from 214 to 301 masl. The lowest point within the DMA is the one where the main pipeline enters into the DMA, and this point was selected for the purpose of incoming flow measurement (the point for the installation of a portable ultrasonic flow meter).

Total water losses are obtained by subtracting the authorized consumption in the DMA (billed and unbilled, which was obtained by reading all water meters in the DMA at the start and at the end of measurements) from the total water quantity entering into the DMA. During the meter reading campaign in the DMA, the number of unmetered connections was checked and their water consumption during the reference period was properly estimated. The incoming water quantity was obtained by permanent flow metering at the entrance pipe into DMA during a period of seven consecutive days.

The real losses are derived from the measurement of the minimum night flow. The real losses are evaluated by subtracting the actual night water consumption in the DMA from the measured minimum night flow rate and by multiplying the resulting volume by the Night/Day factor.

Night consumption is metered by a control measurement of a certain number of randomly selected samples concerning four defined consumer categories in the zone.

- private houses – 10 samples,
- residential buildings – 3 samples,

- small-scale economy – 7 samples,
- special consumers – 2 samples.

The measurement was carried out by reading the consumers' water meters at 24:00 and 5:00 a.m., after which average hourly water consumption in the consumers' installations was obtained.

During the sampling selection of consumers, due account was taken of those who possessed newer water meters. Based on the results obtained, the night water consumption in the DMA was determined.

This measurement activity lasted for one whole week, and during the same period, in addition to the hydraulic measurement, a detailed visit to and reading of all consumers' water meters was done.

All data collected (obtained by measurement or received from ViK) were processed by CalcuLEAKator program, and the water balance components parameters are derived and presented in Figure 9.1.3.

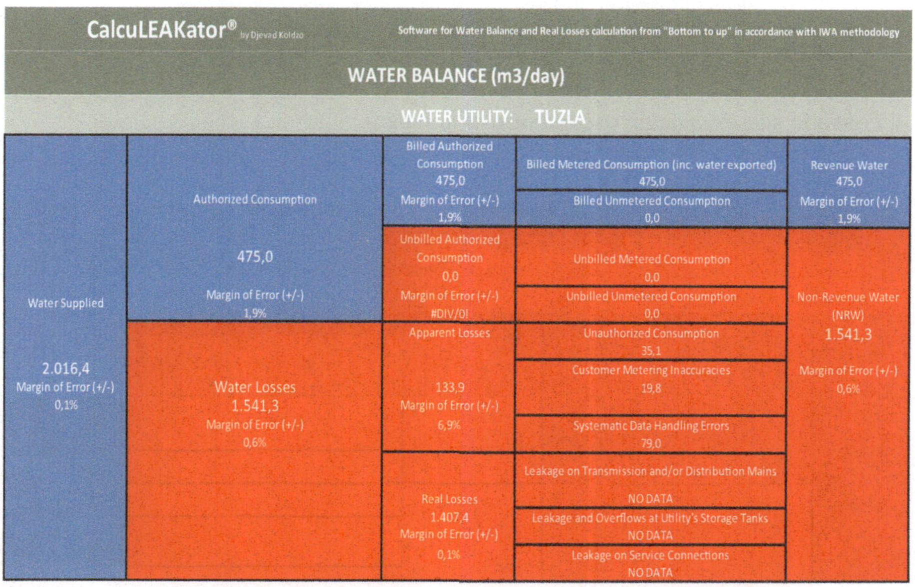

Figure 9.1.3 IWA Water Balance for Tuzla town in CalcuLEAKator 3.0 software (*Source*: Đevad Koldžo).

After Water Balance and Performance Indicators calculation, ILI was found to be 27.4. Thus, this zone is positioned within the Category D, which is the last one of the set of categories (where for developing countries, ILI is higher than 16). Based on the obtained ILI of 27.4, a general conclusion was made that the water losses are very high, and that the implementation of a water loss reduction programme is an imperative and priority.

After the first round of measurements and water balance calculation, sound leak detection was conducted. During the detection process, 10 underground leaks and 19 illegal connections were detected. Excavation of the biggest detected leak (ca 5 l/s) is presented in the Figure 9.1.4.

The ViK Repair Service repaired all of the detected leaks within the shortest possible time, after which the control hydraulic measurement and water balance calculation were carried out at the same measurement points as in the previous phase. The goal of these activities was to evaluate the savings effectuated (if any) from these water leak repairs. Results of second Water balance based on the measurement after leaks repair is presented in the Figure 9.1.5.

Figure 9.1.4 Leakage site excavation (*Source*: Đevad Koldžo).

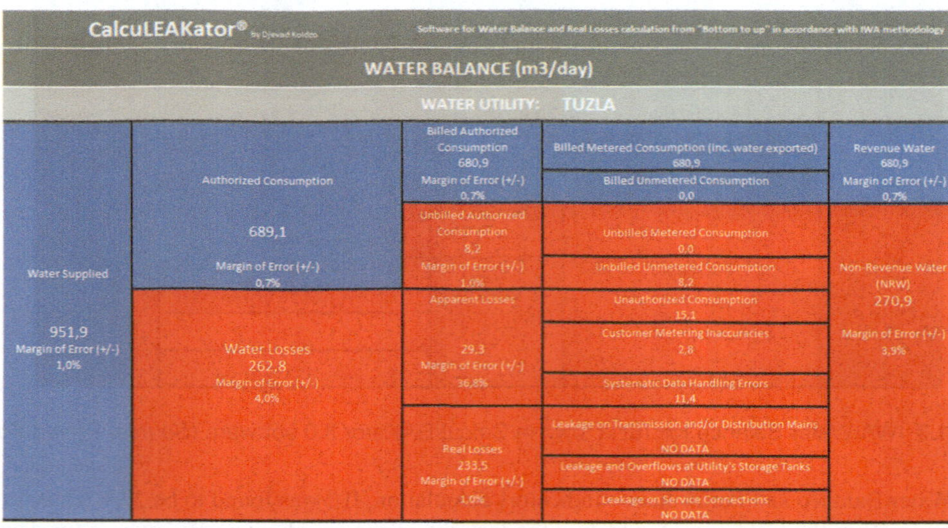

Figure 9.1.5 IWA Water Balance for Tuzla town in CalcuLEAKator 3.0 software (*Source*: Đevad Koldžo).

During the second round of measurements the minimum night flow rate was registered at around 2:00 a.m. and it was 3.6 l/s, which is an expected decrease of 14.3 l/s resulting from the repairs of 10 detected leaks.

All key performance indicators have been improved, and the ILI has decreased from the earlier recorded 27.4 to only 3.9, thus classifying the DMA into Category A. Based on the IWA recommendations for water supply systems in the developing countries, the conclusion concerning the above mentioned is as follows: further reduction of losses may not be cost effective unless there is a shortage of water. It is necessary to identify cost effective improvements through a detailed analysis.

The water loss reduction action, carried out in the Tuzla DMA, resulted in the water loss reduction to such an extent that at the beginning of the project the system management quality in this zone was classified in the worst category (Category D), in line with the IWA recommendations, while at the end of the project, it was classified within the best category (Category A).

9.1.4 CONCLUSION

After the detection of 19 illegal connections within the DMA and 10 underground leaks, out of which three were very large, the conditions were created for this zone that any further leak detection would not be cost effective. Achieved savings shown through values of key Performance Indicators, at the beginning and at the end of project are presented in the Figure 9.1.6.

PERFORMANCE INDICATORS	December 2012.	April 2013.	SAVINGS/ IMPROVEMENTS ACHIEVED
LEVEL OF SERVICE			
Minimal Flow (l/s)	17,9	3,6	14,3
Average Flow (l/s)	23,3	11,0	12,3
Maximal Prssure (bar)	3,9	4,1	-0,2
Average Pressure (bar)	3,8	4,0	-0,3
Authorised Connections Number	973	1080	-107,0
Night Consumption and Wastage (l/s)	0,6	0,6	0,0
Authorised Consumption (l/s)	5,5	8,0	-2,5
APPARENT LOSSES PERFORMANCE INDICATORS			
Apparent Losses in regards to Authorised cons. (%)	28,2%	4,3%	0,2
Liters per connection per day	137,6	27,1	110,5
Unauthorised Connections Number	44	0	44
LOSSES PARAMETERS			
Leakage Exponent	0,77	0,774	0,0
Night to Day Factor	0,97	0,984	0,0
Non Revenue Water (NRW) (m3/day)	1541,3	262,8	1278,6
VOLUME OF REAL LOSSES			
Real Losses in regards to Water Supplied. (%)	69,8%	24,5%	45,3%
Current Annual Real Losses (CARL) (m3/year)	513707,7	85210,5	428497,2
Unavoidable Annual Real losses (m3/year)	18755,7	21956,2	-3200,4
REAL LOSSES PERFORMANCE INDICATORS			
Infrastructure Leakage Index (ILI)	27,4	3,9	23,5
Liters per connection per day	1446,5	216,2	1230,3
Liters per connection per day per metter pressure	38,4	5,4	33,1
m3 per km mains per hours	4,8	0,8	4,0
PERFORMANCE GROUP	D	A	
FINANCIAL PREFORMANCE INDICATORS			
Unbilled Metered Consumption (Euro /year)	0	0	0
Unbilled Unmetered Consumption (Euro /year)	0,0	1857,8	1740,0
Real Losses (Euro /year)	64804,8	10686,6	54118,2
Apparent Losses (Euro /year)	30423,1	6869,6	23553,5
TOTAL VALUE OF NRW (Euro /year)	95227,9	19414,1	75813,9
Volume of NRW expressed in % of System Input Volume	38,5%	14,3%	24,2%
Value of NRW expressed in % of Annual Operating Cost	1,3%	0,3%	1,0%
Liters per connection per day (Euro /year)	797,7	124,3	673,5

Figure 9.1.6 Performance indicators for Tuzla town (*Source*: Đevad Koldžo).

All further activities within the DMA that are carried out by ViK will concern monitoring, and will be implemented through a control flow measurement at the inflow points into the zone.

The high level of non-revenue water identified after the first water balance was a result of several different causes.

Ten leaks were not detected earlier; regardless of the ViK having quality staff for sound leak detection, it still lacks a consistent measurement strategy. The measurement strategy is supposed to guide the leak detection team in areas with recorded high night flow levels.

Considering that VIK Tuzla possesses its own calibration station, it was expected that all water meters in the DMA are in line with the legal obligations. Evidence that this is the case is that during the sound leak detection campaign, no leaks or irregularities with the house connections in the DMA were detected. This situation is very rare in BiH utilities.

After the final report, KfW bank, encouraged by the results, made decision to extend the project for another DMA, considering also the option to extend it to the whole water supply system in Tuzla.

This project also proved that application of the program CalcuLEAKator has enabled quick and accurate data processing, showing that the program is applicable for all projects where calculation of Water Balance is done by a 'bottom-up' approach. Program CalcuLEAKator can be downloaded free of charge at: www.waterloss.com.ba.

REFERENCES

Koldžo Đ. (2004). *Priručnik za efikasno mjerenje i otkrivanje gubitaka u vodovodnim sistemima* (*Manual for Efficient Measurement and Water Loss Detection*), Bosnia and Herzegovina, pp. 31–76.

Koldžo Đ. and Vučijak B. (2013). Testing innovative software tool CalcuLEAKator for water balance evaluation and water loss reduction in Tuzla project, *6th International Water Loss Conference*, Sofia (Bulgaria).

Vučijak B., Čerič A. and Koldžo, Đ. (2013). Key operational performance indicators for water losses in water utilities in Bosnia and Herzegovina, *International Science Conference*: *Reporting for Sustainability in Budva* (Montenegro).

Chapter 9.2

Large scale water loss management improvement programme in Zagreb (Croatia) water distribution network

Kristijan Iličić, Jurica Kovač, Vlado Herceg, Berislav Bohatka, Mario Medven and Ivan Grdenić

9.2.1 INTRODUCTION

Water utility Zagreb (3000 km network) distributes water for the city of Zagreb, capital of Croatia, with 850,000 inhabitants. The utility started its own water loss programme with the development of advanced hydraulic mathematic modelling in the mid-90s, but lost some momentum until recently when the IWA methodology was implemented in 2008 (featuring pilot zone tests of advanced pressure control application and system analysis). Last year it started a comprehensive water loss control program and the main elements are presented in this paper.

Sustainability and optimization of the existing water supply system are based on the establishment of its oversight and management, which is a strategic goal of the water supply and drainage utility. The first step will be achieved through implementation of the Water Leakage Management Program (phase 1) which includes the installation of flow meters on the transportation mains with which the system will be divided into several major water supply metering zones. With the construction of metering chambers and use of mainly insertion flow meters, the system is divided into 20 large zones with a range of inlet flows between 40 and 700 l/s.

The aim is to establish monitoring of consumption and water losses, so that the data via remote data transmission system becomes available in the dispatching centre. This part of the project is financed by the funds of the European Bank for Reconstruction and Development (EBRD).

The next step involves the further division of the larger areas (zones) into smaller District Metered Areas – DMAs, and some of them will be also be Pressure Management Areas – PMAs, in such a way that the average consumption in each zone is approximately 40 l/s of day inflow. Also, the size of the DMAs will allow fast and efficient Active Leakage Control activities (seeking and detecting leakages, apparent losses and unauthorized consumption).

Dedicated programme implementation was initiated with the aim to reduce water losses in the 2013 by 4.6 million m^3 compared to the previous 2012. The dynamics of establishing control of water supply zones and introduction of measurement and analysis will determine the pace and priorities regarding leakage detection (and reduction due to pressure reduction), repair and rehabilitation. The realization of the first part of the project involves the formation of 13 DMA zones, which is in the final stage right now, and in 2014 further expansion is planned, with 15 new DMAs.

9.2.2 DMAs AND PMAs – STARTING POINTS

Zoning is implemented with the goal of obtaining a satisfactory level of consumption, which allows the conduction of active leakage control (ALC). With the zoning as fine as possible, the number of service connections in one zone is optimized, which results in better losses monitoring and quicker pipeline rupture detection. During the process of zoning, it is necessary to maintain the ratio between the costs arising from the construction of a system for monitoring and managing losses, which implies the installation of the measuring equipment, and cost savings that result from the introduction of such a system, so that the cost of construction and equipment does not exceed the effects of savings.

The long time goal is to establish complete control over the 3000 km of water supply system network in the city of Zagreb by the year 2018, with an emphasis on locating and eliminating losses. Partitioning the system into individual zones of control (DMAs) has already started and systematic measures to achieve controlled pressure are being introduced (Iličić, 2012).

When defining DMAs, zone limits and boundary valves must be determined and permanently closed and points of entry into the zones and critical points of pressure monitoring in the zones must be defined. Also, the newly established water network dead ends, which are the result of zone boundaries, must be identified and included into the hydrant flushing programme. Zones are formed so that the night flows that provide a satisfactory estimate of unreported losses and pipeline ruptures can be properly monitored while ensuring sufficient quantity and quality of water delivered to consumers.

In cases when, according to hydraulic analysis and initial measurements, pressure management inside DMAs is possible and required (constant or dynamic pressure regulation), zones become PMAs (Pressure Management Areas). The forming of PMAs implies installation of pressure regulators (pressure control valves) in ports of entry, which depending on consumption within the zones, regulate the valves at the entrance to the zones. In this way, uniform pressure distribution in time and space within the zones is provided.

9.2.3 DESIGN OF DMA/PMA

The realization of DMA/PMA zones begins with a preliminary determination of possible zones within the water supply system, during which the following parameters were taken into account:

- Possibility of exclusion of tall buildings from the zone, because these facilities prevent pressure reduction without affecting the quantity of water delivered to consumers,
- The intensity of water flow through the pipelines after reducing the number of entrances to the zone, in order to ensure satisfactory water quality,
- The significance of water losses due to leakage of pipelines, in order to give priority to critical areas of the water supply system,
- The possibility of isolating the zone from the rest of the system, in order to ensure the accuracy of measured data,
- Critical pressure values, which are the cause of system failures and excessive leaks into the ground,
- Urban positioning of the ports of entrance and the ports of exit in the zone, and the design of the valve chamber with associated instrumentation and control equipment for the measurement and control of flow and pressure,
- Asignment of the pressure measurement points and installation of pressure gauges within the boundaries of the zone (Iličić, 2009).

Our project on which the implementation of the DMA/PMA zone is based contained:

- The exact location of the new valve chambers in which the instrumentation and control equipment will be located,
- The position of the newly established water network dead ends included into the hydrant flushing programme,
- A structural and mechanical project of the valve chamber; the size of the chamber depends on the dimension of the pipeline,
- The means of powering the instrumentation and control equipment, with an emphasis on:
  - suggestion of the means of powering
  - prediction of the necessary additional equipment whose position should be defined in the nearby area
  - obtaining general approval of the other utility companies in whose area of jurisdiction construction of new valve chambers will be carried out (Electric company, Gasworks, etc. …),
- A mathematical model of the DMA/PMA and hydraulic analysis of the zone based on the mathematical model, from which the conclusions of the pressure distribution within the zone and the state of water quality within the zone was analysed.

An example of a wiring diagram of instrumentation and control equipment installed inside the metering and regulating inlet chamber. (Figure 9.2.1).

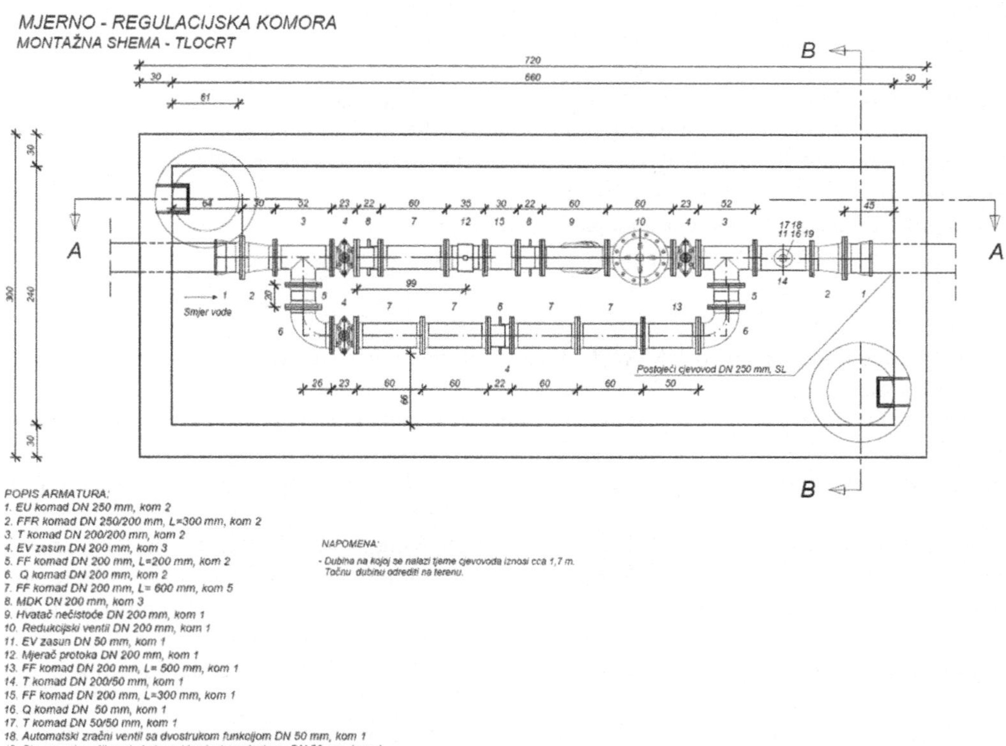

Figure 9.2.1 Measuring and pressure control chamber schematic design (*Source*: Water Supply and Sewage Ltd, Zagreb).

To confirm the quality of the assumed parameters of the defined DMA/PMA zone, measurements of flow and pressure were carried out with the help of mobile measuring and data logging instruments. These measurements were firstly conducted in a situation of a completely opened system, and then followed by a situation where the system is supplied exclusively through the inlet pipeline(s) at the defined point(s) of entry.

9.2.3.1 Detailed design included

- Analysis of the current and the prediction of the future hydraulic conditions within the DMA/PMA,
- Structural project of the point of entry chamber(s), with all static calculations and designs of reinforcement and formwork included,
- Mechanical design (installation works) in which the mounting scheme for each metering point was displayed separately,
- Electrical project with detailed installations of the power supply of instrumentation and control equipment, automatic pressure control, equipment for data transmission, and so on,
- Cost estimates.

The project includes a procedure for testing and commissioning the system for regulating the pressure described in stages below:

- Closing all boundary valves on the edges of the pressure regulating area,
- Supplying the area only through the pressure control valves,
- Initial setup of the pressure control valve and automatics,
- Installation of flow meters and pressure gauges for the analysis of pressure in the area,
- Work check and fine-tuning of the pressure control valve and automatics,
- Continuous monitoring and analysis of measurements in the pressure regulating area,
- Determining the effects of pressure regulation,
- Putting the DMA/PMA into continuous operation.

9.2.4 IMPLEMENTATION

View of the planned and executed DMA/PMA zones in the water supply system of the city of Zagreb 2008–2013 (Figure 9.2.2).

View of the DMA/PMA zones realized during 2013 with entering points; measuring/regulating chambers (Figure 9.2.3).

A DMA/PMA zone with marked peripheral chambers in which boundary valves will be closed (Figure 9.2.4).

Peripheral chambers of the defined DMA zone in which boundary valves will be closed are marked on the Geographic Information System (GIS) map. The closing of the boundary valves is conducted in order to isolate the DMA/PMA zone from the rest of the water supply system, that is, to establish the water inlet to the zone through the monitored point of entry in the new valve chamber.

An example of a chamber during construction. Ready-made industrial metal formwork is used for concreting (Figure 9.2.5).

An example of the installation of already combined instrumentation and control equipment inside the inlet chamber with a crane (Figure 9.2.6).

An example of a waterproofing coating of a constructed chamber, just before the final rehabilitation of the public surface on which it was built (Figure 9.2.7).

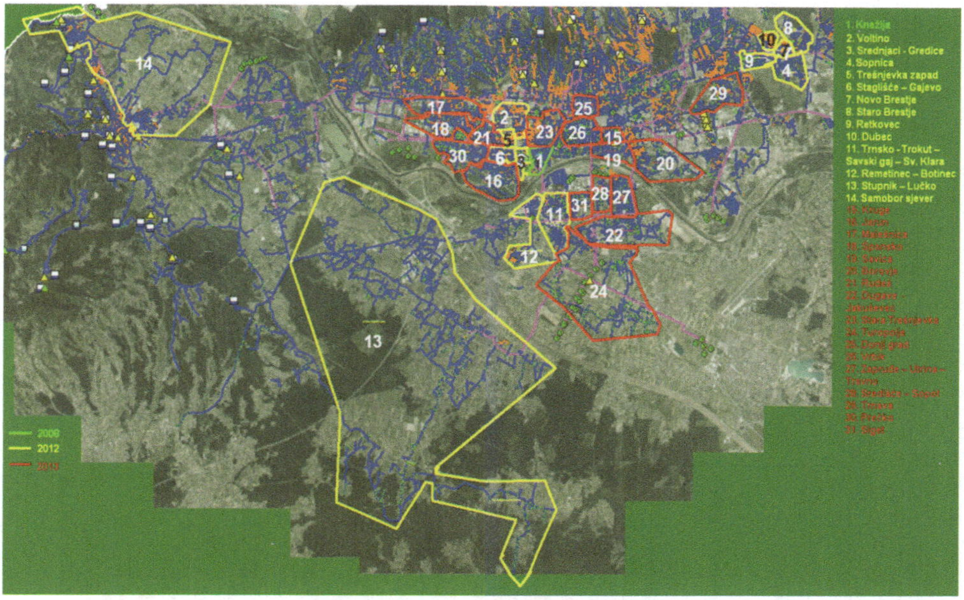

Figure 9.2.2 Implementation of DMA zones in Zagreb (*Source*: Water Supply and Sewage Ltd, Zagreb).

Figure 9.2.3 DMA zones in Zagreb with measuring points (*Source*: Water Supply and Sewage Ltd, Zagreb).

Figure 9.2.4 Boundary valves in one DMA (*Source*: Water Supply and Sewage Ltd, Zagreb).

Figure 9.2.5 Measuring chamber for DMA – construction (*Source*: Water Supply and Sewage Ltd, Zagreb).

Figure 9.2.6 Elements installation in chamber (*Source*: Water Supply and Sewage Ltd, Zagreb).

Figure 9.2.7 Chamber final works. (*Source*: Water Supply and Sewage Ltd, Zagreb).

9.2.5 RESULTS

Implementation (construction works and equipment installation) of DMA/PMAs lasted during 2013 and in the second half of the year we initiated functioning of pressure control in PMAs. Following initial measurements in zones, an active leakage control campaign was initiated, which becomes a continuous activity with the aim to reduce all real losses and keep them under control. These activities will result in further savings of water.

Reduction in pressure resulted in reduction of losses in particular zones which is presented in Table 9.2.1 (*Note:* in zone 'Tresnjevka zapad' also conducted leak detection and repair which resulted in additional savings).

Table 9.2.1 PMA zones with results in savings.

No	DMA/PMA ZONE	Inlet press. before		Inlet press. after		Flow before		Flow after		Leakage before	Leakage after	Savings	
		min. (bar)	max. (bar)	min. (bar)	max. (bar)	min (l/s)	max (l/s)	min (l/s)	max (l/s)	m3/day	m3/day	m3/day	m3/year
1	DMA/PMA VOLTINO	6,41	7,07	5,45	5,45	33	70	25	65	2.614	2.160	454	165.564
2	DMA/PMA TREŠNJEVKA ZAPAD	4,50	6,53	5,00	5,00	18	24	7	15	1.426	605	821	299.592
3	DMA/PMA STAGLIŠĆE-GAJEVO	5,21	6,7	5,1	5,1	12	37	5	31	950	432	518	189.216
4	DMA/PMA SREDNJACI-GREDICE	5,35	6,94	5,33	5,33	19	52	10	49	1.505	864	641	233.892
5	DMA/PMA REMETINEC-BOTINEC	6,59	7,43	5,5	5,5	12	44	10,5	39	950	907	43	15.768
6	DMA/PMA TRNSKO-TROKUT-SAVSKI GAJ-SV. KLARA	7,07	7,65	5,37	5,37	14	40	4	37	1.109	346	763	278.568
7	DMA/PMA RETKOVEC	5,6	6,4	5,1	5,1	6	17	4	15	475	346	130	47.304
											Total m3/year		1.229.904

Source: Water Supply and Sewage Ltd, Zagreb.

The presented measures, and improving the efficiency of the maintenance activities (implemented as a separate programme of improvements in the utility), resulted in reduction of real losses by 3.3 million m^3 in 2013, compared with 2012 (reduced volume of real losses by 6.49%). The main figures of the water balance are presented in the Table 9.2.2 (volumes in m^3).

Table 9.2.2 Water balance figures for 2012 and 2013.

Year	SIV	RW	NRW		RL
2012	120.710.125	59.932.216	60.777.909	50,35%	51.439
2013	114.534.538	57.532.383	57.002.155	49,77%	48.101
Change	6.175.587	2.399.833	3.775.754	0,58	3.338
	5,12%	4,00%	6,21%		6,49%

Source: Water Supply and Sewage Ltd, Zagreb.

Interesting to notice is that NRW in volumes has significant reduction between 2 years but presented as % from SIV (System input volume) change is minor (from 50,37% to 49,77%) – change in percentages is small due to reduction in RW (Revenue water). This shows clearly that NRW in percentages should not be used as the main indicator in a water loss reduction strategy (Zagreb Holding Ltd, 2012).

Also we have calculation of IWA performance indicators that are now recognized as relevant information for future planning of our programmes and expected results. PIs from program CheckCalcs with main indicators for 2013 are presented in Table 9.2.3.

Table 9.2.3 Performance sheet shows main PI for Zagreb utility.

'LEAKS' Suite of LEAKAGE EVALUATION and ASSESSMENT KNOW-HOW SOFTWARE						
CheckCalcs - a free software for identifying Leakage and Pressure Management Opportunities						
CheckCalcs	Developed	Version 3a	29th Aug 2010	Europe	EUR.0001	© ILMSS Ltd
THIS WORKSHEET SHOWS IWA BEST PRACTICE PERFORMANCE INDICATORS FOR NRW, APPARENT LOSSES AND REAL LOSSES						
Colour coding	Data entry	Calculated Values	Default Values	Data from another Worksheet		

Type of PI ↓	European Master Copy	eb water utility Cr 01.01.2013	to	01.01.2014
Operational management of Non Revenue Water	Non Revenue Water (NRW)		1643,9	litres/service connection/day
	Unbilled Authorised Consumption		83,4	litres/service connection/day
	Apparent Losses		11,1%	% of Billed Metered Consumption
	Real Losses	Best PI for Metric Benchmarking	15,68	Infrastructure Leakage Index ILI
		Best PI for Process Benchmarking	1387,2	litres/service connection/day
Financial	Non Revenue Water (NRW)		49,8%	of System Input Volume
			51,2%	of Water Supplied

Source: CheckCalcs software, Western Balkan version (2010).

9.2.6 CONCLUSION

Since the implementation of the strategic goal of 10 DMA zones in one year (in 2013 we succeeded implementing 13 zones), the project activities have been initiated for further 17 zone that will start with implementation later this year. DMA design process is defined by 3 stages; modelling (with use of GIS, Epanet and visual inspections), test measurements (with mobile flow and pressure meters) and final design preparation (including measuring/pressure control chambers). By the end of 2014 we have a plan to establish 30 DMAs (many of them PMAs) in our network.

All activities are conducted exclusively through internal human and material resources (construction and installation works were done by external contractors), and represent a major breakthrough in current management and maintenance practice under the jurisdiction of Branch Water Supply and Drainage. Further maintenance, rehabilitation, reconstruction and replacement of selected pipelines will be outcomes from the knowledge that follows the analysis and management of DMA zones.

REFERENCES

Iličić, K. (2012). Short-term measures measures for reduction of water losses in water supply systems Zagreb, *Professional Business Conference Current Problems in Water Supply and Drainage*, HGVIK, Brač.

Iličić K. and Kovač J. (2009). Systematic approach in analysis of pressure/burst frequency relationship, *WaterLoss 2009*, Cape Town.

Zagreb Holding Ltd. (2012). According to Internal data, Subsidiaries Water Supply and Drainage, K. Iličić: Report on the Implementation of Measures to Reduce Water Losses in 2013 g, Zagreb.

Chapter 9.3

Analysis of the water losses from the aspect of central water meters in objects of collective housing

Bajči Angela and Radovanov Milenko

9.3.1 INTRODUCTION

Municipality of Vrbas is located in the central part of Bačka, in the territory of Vojvodina in Serbia and it consists of the town of Vrbas and five inhabited places: Kucura, Zmajevo, Bačko Dobro Polje, Savino Selo and Ravno Selo. The existing drinking water supply solution regarding the municipality of Vrbas is based on six independent water supply systems, that is, each of the inhabited places gets its water supply through its own local water supply systems from local sources. Maintenance of these water supply systems together with the distribution network is done within the framework of Public Utility Company Standard, Waterworks and Sewage Working Unit from Vrbas, while the maintenance of the internal installation of objects in individual and collective housing are performed by the proprietor of the object, that is, the tenant.

The municipality of Vrbas, according to the last census from 2002, numbers a total of 46,423 inhabitants, while the town of Vrbas has 26,248. At the level of the municipality of Vrbas, about 100% of the inhabitants are connected to the town water supply network. Of the total number of connections, about 95% are connections for households, while the remaining 5% are connections for legal entities – business premises, business objects, and so on. Bigger industrial objects have their own wells.

In all inhabited places belonging to the Municipality of Vrbas, water delivered to citizens in individual housing is calculated on the basis of the recorded consumption using water meters or at a flat-rate (5 m^3 per member of the household monthly), while in the objects of collective housing the consumed water is calculated at a flat-rate for citizens (4 m^3 per member of the household, monthly).

During the year 2004, an internal analysis of the influence of flat-rate users on the total loss of water in the water supply system was done, with special reference to the central water meters in the objects of collective housing. The exiting water meters were calibrated and the central ones out of order were replaced, thus after a certain period there were clear data on the water losses that occur due to flat-rate collection of payments in zones of collective housing. During the year 2011 the German organization GIZ made a study on the water losses in the water supply system of Vrbas, and thus the results of the analysis from 2004 were confirmed, that the greatest losses occur in zones of collective housing with water consumption calculated at the flat-rate.

Unfortunately, even after 8 years and after all the performed analyses and clear findings, we did not come upon support necessary for the solving of the problem that was identified. The principal obstacle in solving of these problems is of the administrative nature that is, reluctance to vote for a municipal decision to calculate the water consumption using central water meters.

Figure 9.3.1 Vrbas (*Source*: Google Earth).

9.3.2 DESCRIPTION OF THE PROBLEM

Of the number of connected households, at the level of the municipality 21.9% are flat-rate users and only in the town of Vrbas there are 31.5% of them. The reason for this is that Vrbas is an inhabited place of the urban type where about 25% of the inhabitants live in collective housing. It should be mentioned that most flat-rate users live in collective housing, which is about 82% of the number of connected flat-rate households.

In the mentioned objects, water consumption is calculated at a flat-rate, and not on the basis of a water meter. In a great number of cases it is not in accordance with actual consumption. This is illustrated by the fact that tenants who live in the objects of collective housing consume, on average, about 6.4 m^3 of water per member of the household per month, while they get invoiced for only about 4.0 m^3 per month, that is, for only about 60% of their actual consumption.

If all the objects of collective housing are taken into consideration, annually they produce a loss amounting to ca 135,000 m^3 of water or ca 50,000 Euros. This means that the buildings produce a loss of water of ca 5.5% in relation to the delivered water or ca 9.0% in relation to the invoiced quantity of water.

All this points to irrational, uncontrolled consumption, waste of water in flats; not maintaining the internal water supply installations in buildings (verticals); out of order internal installations in flats (out of order water tanks, leaking taps, etc.). Consumers do take care of water consumption largely because they do not bear the expenses for the actual water they consume. On the other hand these expenses go directly to the expense of the supplier (PUC Standard). Another important problem should be mentioned, and that is, in the objects of collective housing there are no tenant councils whose task is to solve various problems that occur in individual buildings.

Figure 9.3.2 Water meters testing facility in Vrbas (*Source*: Water Loss Audit report, Kovac, 2010).

9.3.2.1 Example problem 1 – leakage in buildings

We have one more evident example of loss of the supplied water due to non-existence of the central water meter in the collective housing. In street Gustav Krklec no. 36 with 18 flats and 35 tenants, the tenants consume monthly on average about 24 m^3 of water per member of the household, while only 4 m^3 are invoiced, which means there is a loss of water of ca 3425 Euros/year.

Having obtained this datum regarding the great water losses, the Waterworks and Sewage Working Unit sent a request to tenants in street Gustav Krklec, no. 36 to repair damage on the internal installation, however without results. Due to great losses, the damage was repaired by the Waterworks and Sewage Working Unit – but only partly (the damage was in the cellar of the building; water drained into the sewage and did not appear on the surface). However the damage still exists somewhere, since the tenants now consume 15 m^3 per member of the household per month. That datum still points to losses, but the cause is still to be found. The cause may be additional damage to the internal installation in flats or maybe even the unreported members in the households, since the water consumption is calculated at a flat-rate.

9.3.2.2 Example problem 2 – leakage in buildings

The consequences of the reluctance of the local self-government to secure water supply for settled areas in a rational manner are directly felt by the users themselves. This can be illustrated by a case from 2011. In one building in the center of the town, internal water supply installation was damaged. The damage was discovered when the tenants who had flats on the floors above the second floor, complained about the low water pressure at the taps. In addition, the water pressure was so low they could not use the washing machines or warm water from the boilers. After we found the leakage on installation in the common space of the mentioned building, the tenants could not come to an agreement to collect the means to repair the damage. Only after we warned them that we would disconnect the water until they solved the problem did they manage to organize and repair the damage. A considerable quantity of water that had drained during the longstanding damage increased our losses, but if we had the possibility to invoice for the total quantity of water that had been supplied to the mentioned building, we believe that the damage would have been repaired by the proprietors of the flats much sooner, and we would not have had to put pressure on them. It is important to understand again, that the mentioned pressure was exerted only because of the lack of the instruments generally available to water operators (measuring the consumption, billing according to the consumption, etc.), which further worsened the relationship between the company and the consumer. Here can clearly be seen the vicious non-revenue water cycle, which is often described in water loss reduction literature.

For additional illustration the Snapshot Infrastructure Leakage Index (ILI) may be used. It shows how many times the real water losses of the water system are higher than the unavoidable water losses specific for that particular system. In the center of Vrbas, at the zone of the mentioned buildings, the ILI factor is 10.07, while the total ILI factor for Vrbas is 3.38. In the parts of Vrbas where there is a small number of buildings the ILI factor is 1.88, that is, 3.16 (Figures 9.3.3 and 9.3.4).

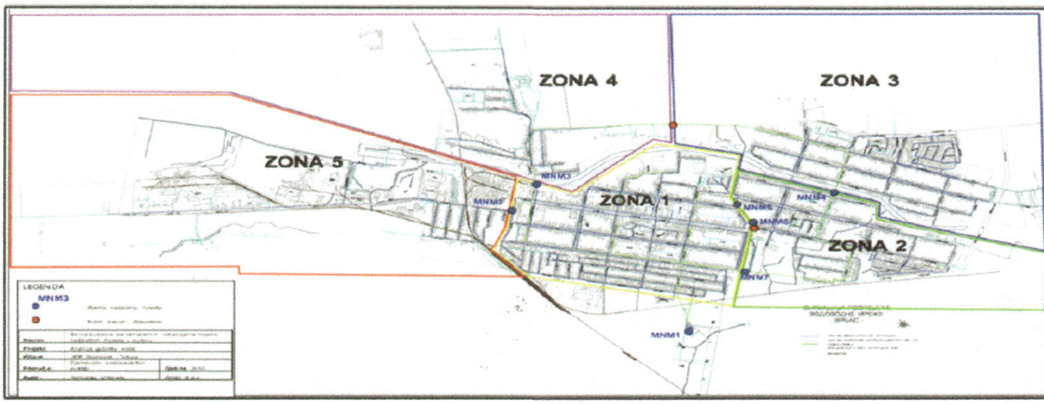

Figure 9.3.3 Vrbas water distribution network with future DMA zones (*Source*: Water Loss Audit report, Kovac, 2010).

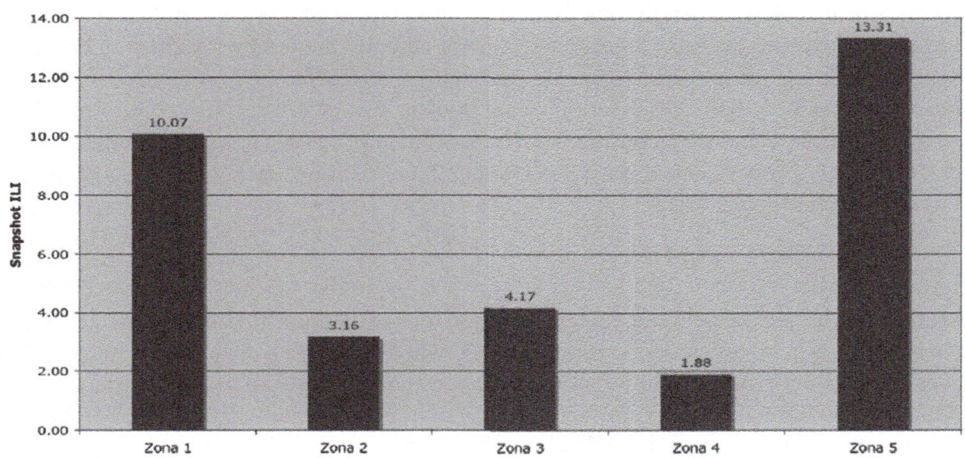

Figure 9.3.4 Snapshot ILI values in DMA zones (*Source*: Water Loss Audit report, Kovac, 2010).

9.3.2.3 Example problem 3 – leakage in buildings

Certainly the most drastic example of to what extent the inadequate action of the local self-government may indirectly induce an unscrupulous relationship of citizens towards such an important resource as water, is in the marginal part of the inhabited place where the Snapshot ILI factor amounts to even 13.31 (Figures 9.3.3 and 9.3.4, zone 5). This means that the real water losses in the water supply system in that part of Vrbas are over 13 times bigger than the unavoidable losses in the particular DMA zone.

In the mentioned part of the town, during the summer time, on the floors higher than the first floor, the citizens have very low water pressure during the greater part of the day. The reason for this is that they themselves or their neighbours use water for watering of the green areas during the whole day and night. At the beginning of 2012 the representatives of the local self-government sent a request to the public utility company to empty septic tanks belonging to one building of collective housing with about thirty flats. Septic tanks were completely filled to the extent that water overflowed to the green areas. When we emptied two connected septic tanks, we noticed that they were recharged with clean water and approximately 60 m^3 of liquid that we drained from the septic tanks were recharged in less than 24 hours.

Due to this, we suspected that there was a possible damage to the water supply installations, and later on we found the leakage in the cellar of the mentioned object. The cellar was full of water that overflowed, found the way to revision chute and thus filled the mentioned septic tanks. Despite all the technical assistance that we gave on behalf of the public utility company and that we were ready to offer to the proprietors of the flats, they did not want to repair the damage. They expected that someone else should do this instead of them. In accordance with the attitude towards water as a resource on the level of local self-government, i.e. that the citizens do not have to pay for water that they actually consume, the citizens have apparently come to the conclusion that they are not obliged to repair the existing damage of installations. For them, the only problem in the given situation represented the sewage water that could not be drained from their flats. Thus, the citizens began to dig open canals around the building, but did not want to repair the damage, although that would have solved the problem of fast septic tank refilling.

The problem was solved when we received assistance from the local self-government to deliver to the tenants a warning from the communal inspection, telling them that water would be turned off if they did not repair the damage. Although the problem in this particular case was solved by the decision of the local self-government at the level of an individual case, in everyday affairs, the public utility company is still forced to exert constant pressure on citizens who are actually guided by the existing decision of the local self-government regarding the low flat-rate payment for water. However, this decision applies only to a number of citizens of Vrbas, thus creating social injustice in society at the expense of those citizens who pay properly for the water they consume. Such an attitude of the local self-government towards the obligations and rights of water consumers is, on one hand, harmful for the consumers, and on the other, for the company which supplies water.

9.3.2.4 Example problem 4 – water meter diameter

The importance of the adequate water meter diameter for the defined flow may also be noticed. For example, in the object in street M. Čobanskog, no 135, where we have 4 flats with a total of 8 tenants, there is a water meter with diameter of 50 mm, which does not measure small water flow, that is, creates apparent water losses. We come across such examples very often, especially in objects where the diameter of the connection is over-dimensioned in relation to the needs.

9.3.3 ANALYSIS

9.3.3.1 User coverage

On the basis of Table 9.3.1 it can be seen that about 100% of population is connected to the town water supply network. Of a total number of connections about 95% of them are household connections while the remaining 5% are legal entity connections (industry has its own wells, that is, own sources of water supply).

Table 9.3.1 Review of the number of water network connections and number of flat rate users based on year 2004.

				2004 (data from 31.12.2004.)							
						Number of flat-rate users out of connected households					
Settlement	Population according to 2002 census	Number of households according to 2002 census	Users per household	**Total number** of connections to the water network. (households.+ legal entities)	**Number of connected households** and % compared to the total no. of connections	**Total** and % wise in comparison to the number of connected households	In condominium housing, % of connected households and % of flat-rate households	In private houses	**Number of connected legal entities** and % out of total number of connections	Number of flat-rate user out of connected legal entities	% of flat-rate users (households + legal entities) compeered to the total number of connections
VRBAS	26.248	14.928	1,75	8.190	7.706 (94, 1%)	2.425 (31, 5%)	1.988 (25,8%) (82,0%)	437	484 (5,9%)	142	31,3 %
B.D.POLJE	3.988	1.165	3,42	1.154	1.107 (95, 9%)	141 (12, 7%)	27 (2,4%) (19,1%)	114	47 (4,1%)	19	13,9 %
ZMAJEVO	4.519	1.404	3,22	1.426	1.371 (96, 1%)	189 (13, 8%)	61 (4,4%) (32,3%)	128	56 (3,9%)	21	14,7 %
KUCURA	4.743	1.644	2,89	1.620	1.567 (96, 7%)	61 (3, 9%)	8 (0,5%) (13,1%)	53	53 (3,3%)	4	4,0 %
RAVNO S.	3.530	1.144	3,09	1.203	1.167 (97, 0%)	102 (8, 7%)	14 (1,2%) (13,7%)	88	36 (3,0%)	15	9,7 %
SAVINO S.	3.395	1.066	3,18	1.003	965 (96, 2%)	124 (12,8%)	36 (3,7%) (29,0%)	88	38 (3,8%)	18	14,2 %
				14.596	**13.883** (95, 1%)	**3.042** (21,9%)	**2.134** (15,4%) (70,2%)	**908**	**714** (4,9%)	**219**	**22,3 %**

Source: Vrbas water utility, Radovanov and Bajci.

Out of all users at the level of the municipality 22.3% are being invoiced according to the flat rate, and not according to the measuring instrument, and only in Vrbas is that percent 31.3%. Out of the number of connected households the percentage of flat-rate users at the level of the municipality is 21.9%, and only in Vrbas is that percentage about 31.5%. This is because Vrbas is the urban type settlement, where about 25% of the population lives in collective housing. In these objects, water consumption is being calculated at a flat rate, and not per water meter, which is not realistic in the great number of cases. Installation of central water meters would provide more realistic calculation and payment of the consumed water would be made possible.

9.3.3.2 Water consumption and flat-rate collection

At the beginning of 2003, the percentage of the flat-rate customers out of the number of connected households was approximately 50%, and at the end of 2004, it was reduced to about 30%. This is first of all accomplished by regular and irregular water meter replacement in the household, i.e. at objects of individual housing. The largest number of flat-rate customers (82%) live in collective housing, and their number would be greatly reduced by the municipal decision that would enable metering of consumed water by central water meters (Table 9.3.2).

On the basis of the existing Rule book on water supply and sewerage applied by PUC Standard it has been foreseen that water consumption for users of housing space is to be measured by water meter, and only exceptionally by estimate. If water consumption is estimated, average consumption of water per capita in an inhabited place for the previous year is to be taken. That estimate was not performed before 2004, and datum on the basis of which the calculation was done earlier was 5.0 m^3 per month per inhabitant for water consumed by flat-rate users in individual housing (houses), and 4.0 m^3 per month per inhabitant, for persons in collective housing (buildings). However, on the basis of the calculation of the average consumption per inhabitant (Tables 9.3.3 and 9.3.4), it can be concluded that the calculation so far is not realistic and that correction of those values should be made. It has been proposed that it should be 5.0 m^3 per month per inhabitant, in objects of collective housing (only in case of the central water meter not being in proper operation), and 6.0 m^3 per month per inhabitant, in objects of individual housing. Thus, it would be in the interest of users to control their own consumption and to report any problems related to water meter operation. Despite the fact that the calculations of the average consumptions are performed annually, the values for the estimate used for flat-rate billing still remain the same as before 2003 (4.0 and 5.0 m^3).

Table 9.3.2 Review of flat rate users compared to the number of household connections.

	2003 (data from 31.3.2003)				2004 (data from 31.12.2004)			
		Number of flat-rate users out of connected households				Number of flat-rate users out of connected households		
Settlement	**Number of connected households**	Total and % wise in comparison to the number of connected households.	In condominium housing, % of connected households and % of flat-rate households	In private houses	**Number of connected households** and % compared to the total no. of connections	Total and % wise in comparison to the number of connected households	In condominium housing, % of connected households and % of flat-rate households	In private houses
VRBAS	7.685	3.857(50,2%)	2.084 (27,1%) (54,0%)	1.773	7.706 (94,1%)	2.425 (31,5%)	1.988 (25,8%) (82,0%)	437
B.D.POLJE	1.101	244 (22,2%)	No data	No data	1.107 (95,9%)	141 (12,7%)	27 (2,4%) (19,1%)	114
ZMAJEVO	1.371	446 (32,5%)	No data	No data	1.371 (96,1%)	189 (13,8%)	61 (4,4%) (32,3%)	128
KUCURA	1.554	131 (8,4%)	No data	No data	1.567 (96,7%)	61 (3,9%)	8 (0,5%) (13,1%)	53
RAVNO S.	1.170	500 (42,7%)	No data	No data	1.167 (97,0%)	102 (8,7%)	14 (1,2%) (13,7%)	88
SAVINO S.	971	599 (61,7%)	No data	No data	965 (96,2%)	124 (12,8%)	36 (3,7%) (29,0%)	88
	13.852	**5.777** (41,7%)			**13.883** (95,1%)	**3.042** (21,9%)	**2.134** (15,4%) (70,2%)	**908**

Source: Vrbas water utility, Radovanov and Bajci.

Table 9.3.3 Average water consumption per capita 2003–2004.

				2003				2004			
Settlement	**Population** according to 2002 census	**Number of house holds** according to 2002 census	Users per house hold	**Invoiced water** in 2003 households + commercial users (m^3)	**Average** per capita m^3/month in 2003	**Delivered water** in 2003 (m^3)	**Average** per capita m^3/month in 2003	**Invoiced water** in 2004 households + commercial users (m^3)	**Average** per capita m^3/month in 2004	**Delivered water** in 2004 (m^3)	**Average** per capita m^3/month in 2004
VRBAS	26.248	14.928	1,75	1.584.932	5,03	2.590.121	8,22	1.486.165	4,72	2.407.745	7,67
B.D.POLJE	3.988	1.165	3,42	264.879	5,53	404.833	8,46	203.359	4,25	303.828	6,35
ZMAJEVO	4.519	1.404	3,22	273.415	5,04	392.966	7,25	250.710	4,62	315.958	5,83
KUCURA	4.743	1.644	2,89	235.286	4,13	338.237	5,94	212.704	3,74	258.433	4,54
RAVNO S.	3.530	1.144	3,09	221.082	5,22	164.092	3,87	188.638	4,45	201.425	4,76
SAVINO S.	3.395	1.066	3,18	204.980	5,03	332.154	8,15	177.299	4,35	229.486	5,63
					4,998 **≈5,00**		7,58 **≈8,00**		4,51 **≈5,00**		6,66 **≈7,00**

Source: Vrbas water utility, Radovanov and Bajci.

Table 9.3.4 Average water consumption per capita 2005–2007.

Settlement	2005				2006				2007			
	Invoiced water in 2005 households + commercial users (m^3)	**Average** per capita m^3/month in 2005	**Delivered water** in 2005 (m^3)	**Average** per capita m^3/month in 2005	**Invoiced water** in 2006 households + commercial users (m^3)	**Average** per capita m^3/month in 2006	**Delivered water** in 2006 (m^3)	**Average** per capita m^3/month in 2006	**Invoiced water** in 2007 households + commercial users (m^3)	**Average** per capita m^3/month in 2007	**Delivered water** in 2007 (m^3)	**Average** per capita m^3/month in 2007
VRBAS	1.428.201	4,53	2.244.517	7,13	1.512.450	4,80	2.279.305	7,24	1.512.098	4,80	2.167.462	6,88
B.D.POLJE	211.959	4,43	301.968	6,31	215.924	4,51	289.756	6,05	236.139	4,93	296.739	6,20
ZMAJEVO	232.306	4,28	263.064	4,85	226.708	4,18	279.000	5,14	235.314	4,34	274.626	5,06
KUCURA	211.100	3,71	266.830	4,69	216.408	3,80	240.715	4,23	232.796	4,09	274.660	4,82
RAVNO S.	171.264	4,04	200.209	4,73	177.342	4,19	209.506	4,95	201.413	4,75	227.795	5,38
SAVINO S.	176.309	4,33	195.245	4,79	191.383	4,70	215.326	5,29	194.822	4,78	220.610	5,42
		4,36 **≈5,00**		6,23 **≈6,00**		4,56 **≈5,00**		6,31 **≈6,00**		4,69 **≈5,00**		6,21 **≈6,00**

Source: Vrbas water utility, Radovanov and Bajci.

9.3.3.3 Analysis – losses due to flat-rate collection of payment

Of the existing objects of collective housing (177 objects, the objects of the settlements Kolonija Šećerane and Kolonija Kudeljare have not been included – about 35 more objects), 89 built in central water meters belonging to the above mentioned objects have been analyzed, that is 50.3%. It has been shown by the analysis that:

Tenants of the mentioned objects consume on average 6.4 m^3 of water per member of the household per month, while only 4.0 m^3 have been invoiced, that is, only 60% of the consumption.

For 89 objects of collective housing 9421 m^3 of water is invoiced monthly per advance payment, instead of 15,070 m^3, that is only 60% is invoiced compared to what has been found out on the basis of installed and controlled central water meters. This means that 5649 m^3 of water has been invoiced less, that is, 37.5% less in relation to the actual quantity of the water consumed.

This at the end of the year brings a loss of about 50,000 euro (12 months × 5649 m^3 = 67,788 m^3; 67.788 m^3 × 30 din = 2.033,640.00 din; 2.033,640.00 din/80.00 = 25,420 euro; about 50% of central water meters have been processed ⇒ 25,420 euro × 2 = 50,840 euro).

Remark: the price of 30.00 din/m^3 is a datum from year 2004 when the exchange rate was 1 euro = 80.00 din ⇒ 1 m^3 of water = 0.37 euro, and the same in 2012 is 46.17 din/m^3, that is with exchange rate 1 euro = 112.00 din ⇒ 1 m^3 of water = 0.41 euro, thus the loss is 55,900 euro).

For example, in the building in the street Gustav Krklec, no. 36 with 18 flats and 38 tenants, consumption on average goes up to 24.03 m^3 of water per tenant per month (on the basis of the recorded state of the built-in central water meter), and that is monthly 913.12 m^3 of water in total, while only 152 m^3 of water are monthly invoiced for the whole building (on the basis of the advance payment). That means that water loss is about 3425 euro annually for the mentioned building only (913.12 – 152.0 = 761.12 m^3 of water monthly; 761.12 m^3 × 12 months = 9133.44 m^3 annually; 9133.44 m^3 × 30.00 din = 274,003.20 din or 3425 euro annually, which in year 2012 amounts to 3765 euro/annually).

The above mentioned indicates: irrational, uncontrolled consumption and waste of water on the part of the users who get charged for consumed water at a flat-rate; failing internal installations (water tank out of order, taps that leak, etc.); lack of maintenance of the internal water supply installations in the buildings (verticals), and points to the importance of having central water meters.

9.3.3.4 Results of the analysis

In the municipality of Vrbas there are 177 objects of collective housing (inhabited places Kolonija Šećerane and Kolonija Kudeljara have not been included, where there are about 35 additional objects) where water consumption should be measured by central water meters that have different diameters (from Ø20 mm to Ø80 mm). Out of the mentioned central water meters 103 pieces have been installed (58.2%), and the remaining 74 pieces need to be installed. For that work 420,000.00 din (3750 €) need to be spent. That is approximately the same amount as the amount of losses made by only one building in the street of Gustav Krklec, no. 36 during one year.

Besides this, every building needs to establish its own tenant council with its own representative who would be obliged to report to the authorized department of the distributor every change concerning the tenants, number of members of the households, changes in the ownership of the flat, and so on. Also, they are to take care about the internal installation of the building, and to solve and repair different damages in due time.

It is also necessary to adopt municipal decisions concerning the changes and supplements to the regulations on water supply and sewerage, and to introduce calculation of water consumption based on the central water meters in the objects of collective housing.

9.3.4 CONCLUSION

In order to solve all the described problems that are connected with water supply in Vrbas it is necessary to make an adequate municipal decision on calculation of the water consumption based on the central water meters, that is, to enable the public utility company to invoice for water according to realistically calculated consumption. Unfortunately, despite continuous demands made by the employees at Waterworks and Sewerage Working Unit, and documented reasons why it should be done (including this paper), instead of adopting the decision, local self-government representatives are trying to solve the problems of water supply in different other ways. In the case of Vrbas decision makers are considering drilling new wells and/or extension of the distributive network, which is clearly the wrong approach frequently used in many other social environments where no respect is given to the important role of a professional approach to solving of water supply problems. Instead, the representatives of the local self-government should make decisions according to their personal conviction, disregarding the demands of the profession, local professionals and recommendations from successful local self-governments, that is, to charge for water according to measured consumption, wherever it is possible, all in order to protect natural resources.

An additional problem is that they try to resort to the mentioned solutions because of frequent complaints from individual citizens that the quantity of water and pressures in the system are inadequate, not understanding that in such a manner the existing problems are only further deepened. On the other hand the citizens who do not have official information on how much water they really consume, or do not know what is the amount they should pay for the water they consume in order to be equal with other citizens in the same community, cannot be good partners in the process of the protection of the environment they live in. Thus, the existing attitude of the local self-government disables constructive discussion between the representatives of the public utility company and the consumers. Therefore, it is not possible to achieve satisfaction for the citizens by the service, despite the unrealistically low prices for the water they consume.

With this paper we have tried to point out how the flat-rate billing of users for the consumed water influences the calculation of losses in the water supply system, and the attitude of citizens towards water, attitude of the public utility company management, as well as the attitude of decision makers in the local self-government.

Losses that occur have a great influence on the existing water supply system, as every loss induces new losses and new problems. The problem of irrational water consumption in objects of collective housing, noted for almost a decade by the technical service that has been maintaining the water supply system, in time, occurs in different forms, but then the solving of the problem is more expensive for the users themselves, and especially for our company.

Until we understand that water is an existential necessity of a society, and that it must not be a political category, we shall have problems in securing adequate quantities of water of adequate quality to our users.

REFERENCES

Kovac J. (2010). Water Loss Audit, GIZ project in Serbia, Strengthening Local Self Government.

Radovanov M. and Bajci, A., personal documentation from Vrbas water utility archives.

Chapter 9.4

Water loss management in Novi Sad water utility, Serbia

Zlatko Arvaji and Ikonija Karadžić

9.4.1 INTRODUCTION

Water losses in water supply systems in Serbia range from 20% to 70%. There are no precise data on the national level, or a national strategy of water loss management. Individual examples of good practice are exceptions. Novi Sad is one of them.

As the second largest city in Serbia with a population of 350,000, Novi Sad has a water supply system (Water Novi Sad) capacity of 1500 l/s, 1000 km mains of average age 35 years, 53,000 service connections and an annual volume of NRW_{2011} = 7.4 MIL m^3 (ILI = 5.0). After years of economic and political crisis during the nineties, since the year 2000, several large projects have been implemented which regained the quality of the system almost to the level before the crisis: reconstruction of 14 km mains of the old city center and 4 km of transit pipelines, a 10,000 water meters replacement campaign (20% of total), forming of 14 DMAs and the construction of a new main pump station with frequency pressure regulation. It marked the beginning of active water loss management, and the results gradually became apparent (Figure 9.4.1). Today, applying the IWA methodologies, Water Novi Sad actively manages water losses in order to achieve results that will make it a more effective and sustainable system for the future.

9.4.2 PRESSURE MANAGEMENT

The system is a combination of gravity and direct pumping. It consists of 14 reservoirs and 17 pumping stations with 15 pressure zones. The first zone covers 70% of the total. The average pressure is relatively low (33 m). Pump stations were not equipped with frequency regulation of pressure. In 2009 a new main frequency regulated pump station was built of capacity 1500 l/s. Average pressures were reduced only 10%, but the hydraulic system stability greatly increased. Since then were observed the trends in permanent reduction of water losses (Figure 9.4.1).

The main part of the system (70% of the network) is on flat ground, with a balanced pressure regime and decreased night time pressure by 5 m. The smaller part of the system (30% of the network) supplies the villages on the slopes of Fruska Gora mountain, with average pressure 55 m, and maximum 120 m. Over time, the hydraulic condition of this part of the system made it much more vulnerable in terms of number of failures. Later analysis of DMAs has shown that the worst performance indicators of actual real losses are in this part of the system (Figure 9.4.2). Long-term development plans provided the reconstruction of the pressure zone in order to reduce the average pressure by 15 m. Works are in progress, followed by active leakage control (ALC) and the expectations are that this part should be stabilized to the level of average PIs of the entire water system (ILI ~ 5).

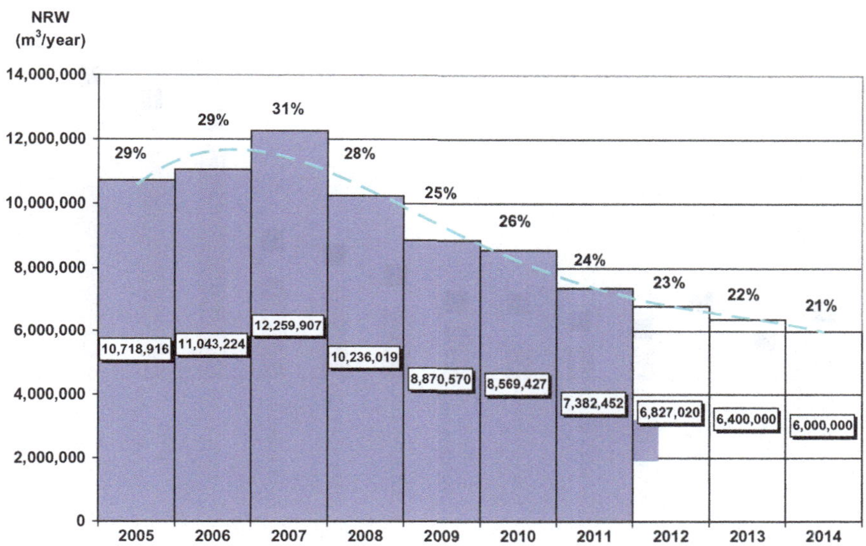

Figure 9.4.1 NRW 2005–2011 with Targets of Strategy 2014 (*Source*: Novi Sad water and sewage utility).

9.4.3 DISTRICT METERED AREAS AND BASIC REAL LOSS PERFORMANCE INDICATORS

Since 2002, there are 14 DMAs that form 50% of the system with continuous SCADA monitoring of flows and pressures. They control the suburb settlements. The remainder is the 15th DMA, an urban mega area (population 200,000) that is in dividing process for new optimal sized DMAs. Since 2009, after the adoption of the IWA methodologies for DMA audits, detailed water losses analysis has been carried out. These results showed the DMAs with the highest key real loss PIs and thus was created a priority list for implementation of the ALC (Figure 9.4.2).

Even though Serbia belongs to the group of developing countries (Low Income Countries) according to the World Bank Institute standards, all water loss analyses in Water Novi Sad are based on more stringent criteria of PI – for developed countries (High Income Countries).

9.4.3.1 Micro metered areas

The specificity of the Water Novi Sad is 250 Micro-Metered Areas (MMA) which were originally intended to control the losses of illegal networks, in the refugee colonies built during the nineties. Each one consists of a main water meter at the junction point to the legal network, own network and services with control water meters – typically poor quality. In total, they represent a 100 km of mains with 10,000 services of the 'temporary registered consumers' (~40 conn./zone). However, due to the undefined maintenance jurisdiction of these zones, through time they became the most vulnerable part of the system. The result was rapid growth of NRW over the years, on 10% of the system.

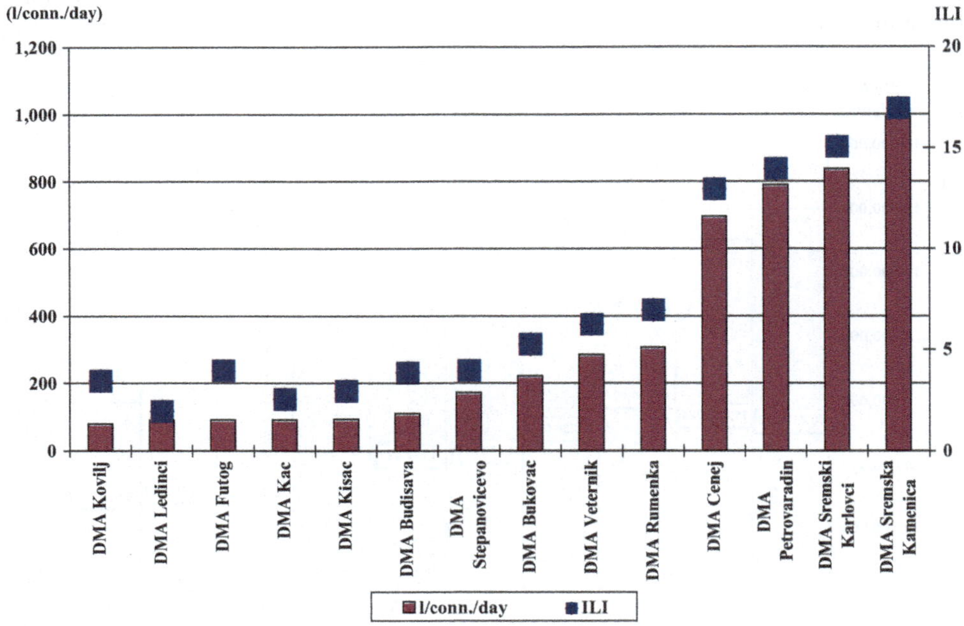

Figure 9.4.2 DMAs Real Loss Performance Indicators 2011 (*Source*: Novi Sad water and sewage utility).

In the period 2009–2011, following the process of the legalization law of buildings and installation, Water Novi Sad has managed to solve most of the leakages of these networks in cooperation with the inspection services. Replacement or installation of the new main water meters and NRW analysis of MMAs helped to obtain relatively easily the information about the numerous leaks which were sometimes even visible. At the same time, an additional inspection of illegal consumers reduced their number. Results were above expected (Figure 9.4.1). The reduction of NRW in 2011 was 1.2 MIL m^3 and was primarily the result of 30 leakage points on the MMAs of illegally built networks, which have been discovered, estimated (~24 l/s = 750,000 m^3/year) and repaired during the year.

9.4.4 ACTIVE LEAKAGE CONTROL AND THE PILOT DMAs

Although the Water Novi Sad from 2000 possessed the equipment for the ALC, it was used only in a passive way (on request). In the meantime the equipment became useless as it became old and was inadequately maintained. During the summer 2011 on the regional level a training and audit was held of the largest pilot DMA, Futog – 30,000 consumers. This intermunicipality project with participants from twelve water systems was guided by a regional IWA-WLSG expert. The whole process, from the network measurements to the software analysis was elaborated. New equipment was purchased (correlator, noise loggers) and ALC initiated on the highest water losses DMA (Cenej – 1500 consumers). Although it is the smallest DMA, ALC investigation was helpful to gain experience and to define steps in systematic work.

It was noted that an incomplete GIS database of services and consumption makes the field work in locating the leakages very hard. The conclusion is that before starting the ALC, all issues to reduce commercial losses must be carried out:

- Replacement of faulty or old water meters.
- Reconstruction of oversized water meters.
- Definition of all objects that are not connected to the water system – potential unauthorized consumption.

After implementation of these activities, a new NRW evaluation of DMAs is required. Avoiding the commercial losses influence in NRW assessment is a priority, otherwise the ALC will be a hard task, especially on a non-metal pipeline DMA.

9.4.5 WATER PRICE POLICY AND WATER METER INACCURACIES

Although the resources of drinking water in Novi Sad are limited, the water price in the past, as well as in the rest of Serbia, was not economically justified. From 2010 the company management supported by the city government changed the water price policy and from 0.35 EUR/m^3 reached the level of 0.84 EUR/m^3 (water + sewerage). This change had a positive effect on demand reduction (10%) and thus the reduction of commercial losses due to meter inaccuracies.

Water Novi Sad has Class A service water meters. A large number of the 53,000 water meters are defective or old meters (40%). The share of commercial losses to total consumption is 14%. This is estimated in accordance with the recommendations of the IWA (Kingdom et al., 2006). NRW components analysis is shown in Figure 9.4.3. The component of unbilled authorized consumption is a negligible volume.

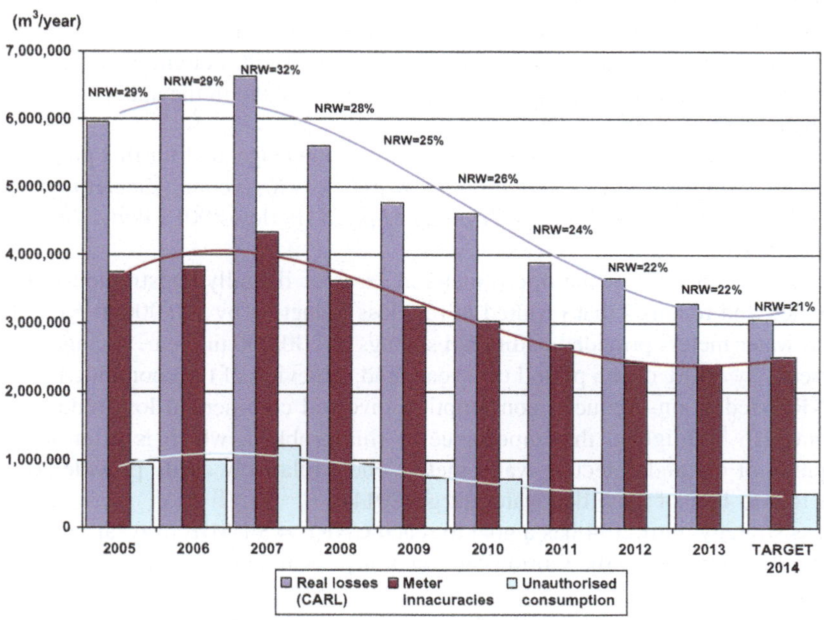

Figure 9.4.3 NRW Components 2005–2011 and Strategy Targets 2014 (*Source*: Novi Sad water and sewage utility).

It is estimated that losses due to meter inaccuracies are a potential source of additional income ~1 mil EUR/year. However, at this moment, the company management is not interested in intensifying water meter replacement, avoiding the further pressure on customer budgets. The present average costs for water and sewerage are 2.6% of the average total household income.

9.4.6 STRATEGY 2014

There has been significant progress in detailed analysis of water loss management and concrete results: NRW_{2007} = 12.3 MIL m^3/year, NRW_{2011} = 7.4 MIL m^3/year. The water loss strategy 2014 defines the target level of NRW_{2014} = 6.0 MIL m^3/year (ILI = 4.0) which will classifie Novi Sad in category B (2 < ILI < 4) in the World Bank Institute classification.

The activities described in the paper should achieve the key target PI_{2014}. After that, a strategy 2019 will be established involving the ELI (Economic Leakage Index) in the final real loss PI definition.

9.4.7 CONCLUSION

In the last 20 years Water Novi Sad went through a period of difficult operation. Intensive investment in the last decade has improved the performance and system reliability. Although there were DMAs and ALC equipment, real loss management was not implemented. By adopting the trends of energy efficiency and sustainable development, water loss management has imposed itself as a new process. In 2010 transparent the IWA methodologies were adopted for analysis and water loss reduction. The process of education has been successfully implemented through the regional workshops and exchange of knowledge and equipment between Water Novi Sad and 12 waterworks utilities (Kovac, 2011). DMA water balancing and the minimum night flow analysis provided answers to the key questions: how much, where and why the water is lost. Measurements and calculations are the most important, but engineers should not run from empirical estimates of unknown values. It is better to make a mistake in the analysis and correct it by the next iteration, than doing nothing.

Also in future activities, it is important to use available knowledge, and for this purpose our engineers regularly use the following literature: *Guidelines for Water Loss Reduction* (Ziegler *et al.* 2011), *Leakage Management and Control – A Best Practice Training Manual* (Farley, 2001), *Water Loss Control* manual (Thornton *et al.*, 2008).

In 2011 through the analysis of the micro-metered areas of illegally constructed water networks, 30 leaks were detected and repaired that resulted in real loss reduction by 750,000 m^3/year. Reconstruction of 100 oversized water meters provided additional savings of 200,000 m^3/year. Change of the water price policy – an increase by 300% over a period of 2 years and achieving of the economic level, had a positive effect on the NRW reduction. Reduced consumption involved commercial loss reduction due to meter inaccuracies, but also highlighted the importance of this problem, which is often underestimated. A replacing campaign of old and defective water meters could relatively easily provide commercial losses reduction from 14% to 10% of the billed water (target 2014).

The water loss strategy until 2014 has a goal to reach CARL = 3.1 MIL m^3/year and the B category of real loss PI (ILI = 4), according to the World Bank Institute classification. Results to date indicate that this is a realistic plan.

The total annual NRW in the period 2007–2011 was reduced by 40%. In 2011 the NRW reduction was 1.2 MIL–m^3. The next period will show how far Novi Sad can go in water loss management, in a low income country environment.

REFERENCES

Farley M. (2001). Leakage Management and Control – A Best Practice Training Manual. World Health Organization, Geneva, Switzerland.

Kingdom B., Liemberger R. and Marin P. (2006). The Challenge of Reducing Non-Revenue Water (NRW). The International Bank for Reconstruction and Development/The World Bank, Washington, DC, USA.

Kovac J. (2011). Water Loss Audit in DMA Futog – Novi Sad, Report, PUC Waterworks and Sewerage, Novi Sad, Serbia.

Thornton J., Sturm R. and Kunkel G. (2008). Water Loss Control, 2nd edn. McGraw-Hill, New York, USA.

Ziegler D. *et al.* (2011). Guidelines for Water Loss Reduction. German Society for International Cooperation (GIZ), Eschborn, Germany.

Chapter 9.5

Continuous acoustic monitoring – evolution of an innovation in the USA

David Hughes

It is well known that leaks on pressure water pipes cause vibration that can be detected by a variety of sensing instruments. Options include continuous acoustic monitoring (CAM), which employs electronic acoustic devices positioned throughout a water system that collect 'noise' data daily and communicate it to the operator. Communication can be by either a mobile pickup or sent through radio or cellular networks back to the host.

Less than ten years ago, CAM devices entered the utility market with some promise. The acoustic devices deployed were connected to valves in valve boxes or to service lines in more protected meter locations about 300 to 500 feet apart as conditions permit. The dynamic of connecting the devices in an advanced meter infrastructure (AMI) introduced the opportunity of listening to a system every day. The technology allowed water utilities to acoustically track the development of leaks. More importantly, the devices had significant success in finding leaks that would otherwise require extended and frequent leak surveying to match.

American Water who pioneered the AMI acoustic monitoring practice for utilities in 2005 (along with Flow Metrix monitors and Hexagram AMI) conducted a three year analysis (2006–2008) of the practice to evaluate the effectiveness of the first CAM system. The reporting of its use in Connellsville, PA, a 5000 customer community with pipes over 100 years old, offered promise but also raised some questions about CAM overall effectiveness in a utility environment. On the positive side, a substantial number of hidden unobserved leaks were found, some of which may have been running for months or possibly years. And after the system was established, subsequent leaks identified before they came to the surface were repaired faster, which led to a substantial initial reduction in water loss. The quick detection reduced the number of emergency repairs (reducing overtime labor costs) and repairing many leaks close to their beginning reduced the average cost of repair from reduced subsurface damage.

The programme pointed to the value of monitoring continuity. After the initial review of the Connellsville system in 2005 and repair of several leaks, a second wave of leaks (some that did not surface) was detected. It is believed that the first thorough review of the system and subsequent repairs may have improved pressure throughout the system to a higher pressure level not seen in the recent past, stressing the network. Essentially, the first leaks repaired had been functioning as system pressure relief valves. This observation should serve as a caution to utilities to be aware that a significant wave of repairs from periodic leak surveys may bring about the rapid development of more leaks that may not be addressed for some time. This trend diminished through 2005 which is termed the 'break in' period. The research conducted with financial support from the Water Research Foundation focused on the subsequent 3 years of acoustic monitoring (2006–2008) after the break-in period.

The research study revealed some limitations of the CAM technique in its initial form though mechanically the system functioned well through the study period. Issues identified included:

- About one third of those leaks that ran 24 hours or more (and should have been detected by overnight monitoring) were not detected by the acoustic monitors.
- Discerning leak noise from other sounds such as heating, cooling and electrical activity was not easy to do. The initial programme relied on constructing a history of sound to identify locations where noise was cyclic: especially in hot and cold weather. Some monitors had background noise that made detection of leaks difficult.

The primary cause of the leaks not detected was directly related to the nature of pipe, pipe leaks and their repair in Connellsville. There was an abundance of old galvanized 2 inch steel pipe and cast iron pipe (4 inch and less) that served as water mains in many small streets. Not surprisingly, the steel pipe was the most common pipe to fail and small cast iron failure was the next most common. Often deterioration in and around the break required removing some portion of the pipe. The method of repair typically involves replacing the removed section of the pipe and replacing it with PVC pipe, secured by two repair clamps. The number of such repairs number in the hundreds in this old pipe network. It is theorized that nearby acoustic monitors are unable to detect leak noise that travels through the multiple types of pipe. The noise is attenuated as it traverses metal pipe and plastic pipe and repair clamps that disrupt pipe continuity. The acoustic monitors had detected leaks from a variety of pipes, including plastic and asbestos cement, but appeared challenged by the pipe transitions.

Figure 9.5.1 Checking a CAM reported leak suspect with field leak noise correlation (*Source*: David Hughes).

The primary drawback with the system was the high number of false positives. Even with some historical clues about how background noise would vary with the time of year, only about 20% of leak investigations proved to find a leak during the study period. Noise detection by the CAM units was time based. An internal clock was set on each unit and noise was detected between 12:30 am and 4:30 am, generally thought to be the quietest time of a day with the unit selecting the quietest 10 minute period that it detected over those 4 hours. After eight years of use and no ability to check time, there is some concern that the time being monitored has drifted to some extent. The manufacturer of the device today (Itron) has elected to have the unit detect sound round the clock. But during the early research period it is believed that time was reasonably well kept by the units.

Despite missing some leaks and the number of false positives, the value of the continuous acoustic monitor was still significant. It should be noted that some leaks by their very nature burst and surface immediately. For the sudden breaks that seem to occur most frequently sometime after the start of the noise detection period, the absence of noise on the night of the burst (form a quiet period prior to the burst) confirmed the suddenness of the failure. The system is fed by PRV units and pressure data collected in selected locations confirmed that maximum pressures frequently occurred in the network in the middle of the night when pressure loss from flow is minimal.

A primary benefit of the research was an improved understanding of the nature of leakage in this small system. As noted earlier, leaks could be tracked acoustically with time, including the time when they began or at least became acoustically significant. The acoustic noise signatures tend to demonstrate that many leaks grow at a very slow rate and consequently do not surface immediately. Many leaks had an elevated sound at the onset that dropped slightly before beginning a slow increase. This initial high noise is attributed to the start of the leak creating more sound as water first breaks through the rough pipe opening that may smooth with water flowing past as well as soil being pushed away from the pipe until a pocket is formed.

Interestingly, over half of the leaks that surface slowly in this system, occur in the last three months of the calendar year. A suspected trigger for the start of many autumn leaks is the sudden changes in water temperature with colder weather, which causes thermal stress on the fragile pipes. (Water is delivered through PRV valves via a local municipal operation from a surface water treatment plant in Connellsville.) It is thought that metal pipe materials attempt to contract in the cold but they are constrained by end connections and pipe cover, causing a tensile stress. Many breaks appear to start when temperatures drop in the fall. In summer, the expansion would cause compression stress which the pipe material has greater capacity to manage.

American Water has added only a few additional acoustic monitoring systems into its other networks, primarily in leaky systems where the cost of water is high. In recent years the confidence of operations staff to find leaks routinely with this approach has waned in Connellsville, owing largely to false positives. Shortly after the American Water experiment began, other leak detection companies pursued the CAM approach to leak detection. But the success of the other products still had similar issues with detecting all leaks and false positive signals. This first wave of the technology was a good start but the value of the method could be enhanced if the vendors improved the technology in one or more of the following ways:

- Significantly reducing false positives. These appear to be most significant when the weather is very hot or very cold and mechanical and electrical equipment is more likely to be in continuous use. In contrast to locations in the Northeast US, a location in northern California has far fewer non-leak noisy locations.
- Providing better information on location of the leak. While leaks found are typically within 500 feet of the CAM unit identifying the leak, investigation work must still be performed (correlators) to find the leak when investigators are sent to the field.

- Providing some improved guidance on the more significant leaks. If there are multiple leak suspects at any one time, some form of prioritization would be useful. To this point, prioritization has been focused on the likelihood that a noise is a leak as opposed to what order of magnitude it may be. Amplitude of leak sound is not a useful criterion. But smaller leaks generally generate a higher frequency so leaks with low frequency could be emphasized.
- Providing a device that is more resilient in the environment. Initial devices located in meter pits had a high degree of failure. Some units located in valve boxes, a harsher environment also had issues.

The Gutermann CAM device was the first to resolve these issues significantly (Figure 9.5.2). Their modifications also sought to address the issue of building a robust detection unit and transmitter. Working with AMI vendor Aclara (formerly Hexagram), the transmitter and CAM units were moved from the meter area or the valve box to be encased in the valve box lid itself. The sensor was hardwired and placed, like other systems onto the valve box nut using a magnet. Gutermann modified the leak detection devices to function like both a leak sensor and a leak correlator. By placing the units close enough together, the units could literally correlate leaks between two adjacent sensors. This not only pinpointed the leak location but served to reduce false positive signals as extraneous noise was far less likely to correlate. In demonstrating this AMI technique, American Water found the system to be successful, though some false positives persisted. However, both the cost of the more sophisticated units and the need to move them closer together altered the economics of the programme and American Water consequently engaged in a 'lift and shift' programme. By periodically moving the units to other parts of a leaky system more value was derived from the units. An Ami system was constructed in two networks not for meter reading but for the express purpose of monitoring for leaks.

American Water continues to test new systems with potential to advance the technology. The latest effort working with Echologics (owned by Mueller who provides the AMI system) may have taken the next step in technology. Echologics has moved the sensor from the meter location or valve box to the inside of the hydrant steamer cap (Figure 9.5.3). This is the same location that Mueller has successfully employed for its AMI mesh network transmitter/repeater. The two functions, acoustic monitor and AMI repeater will soon be available as a single unit.

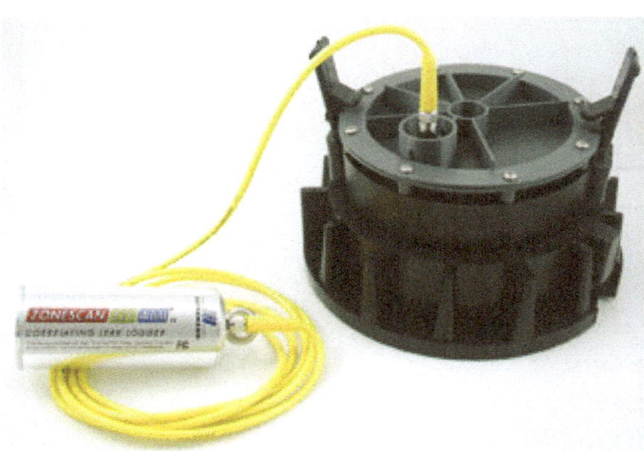

Figure 9.5.2 Gutermann Zone Scan wired to an Aclara transmitter housed in a valve box (*Source*: David Hughes).

Figure 9.5.3 Echologics CAM unit is placed inside the hydrant cap with AMI communications (*Source*: David Hughes).

Prototype acoustic monitoring units tested at Pennsylvania American's Uniontown system and in the City of New Orleans appear to have eliminated the majority of false positives. The units are also capable of correlating (as demonstrated by Gutermann). The units have also been successful in locating leaks where the leak noise had to travel through a combination of metal and plastic pipe.

From the beginning, one feature that has been provided effectively in CAM systems is a data management system 'cloud.' The data once transmitted through an AMI system or in some cases, collected frequently from the field, is brought to the vendor who typically provides a GIS based website where information about the individual sensors can be easily spotted. Systems also have a dashboard identifying primary areas of interest and often prioritize the list of locations to investigate based upon an algorithm that assesses the likelihood that an active noise source is a leak. If this uncertainty (false positives) can be reduced substantially, this prioritization could look to evaluate the possible size of the leak based on its pattern of noise.

The data management display then goes in one of two directions, either providing detail about the latest noise signal or providing a historical display of the noise. The historical display provides a means to help eliminate chronic false positives from active investigation. The Connellsville CAM is the best example in which noise associated with leaks is divided into three groups of frequencies associated with leaks, and displays them over extended time. The three data lines are combined into a summary curve and are taken from the quietest monitored time. It looks at another parameter by looking at difference between the most intense noise in one night versus the quiet time. The expectation is that a leak can become a dominant sound and this parameter will drop as the leak increases. The other type of display is a more detailed examination of the daily record that offers more clues for the experienced leak investigator who can judge by frequency and band width the nature of leaks. The example provided in Figure 9.5.4 shows the normal pattern against a new noise pattern but it can also allow the user to view how the pattern has changed over time by viewing previous days. These systems feature their own algorithms that can interpret and or prioritize the acoustic sound in an effort to focus on the more severe leak or the sound more likely to be a leak.

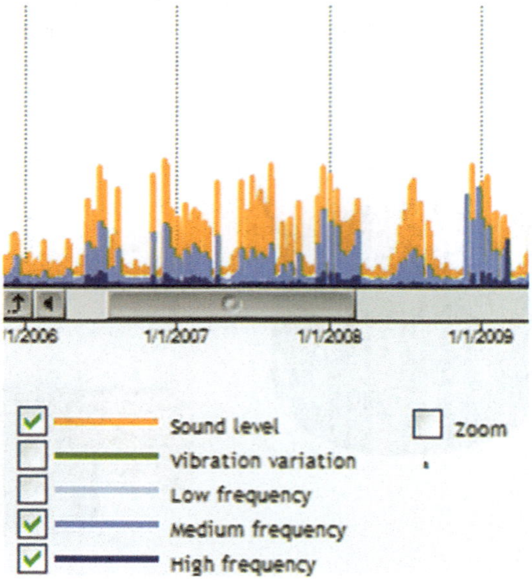

Figure 9.5.4 Five years of data from a Connellsville CAM shows a repetitive pattern of noise from non-leak sources varying largely as a function of weather (*Source*: David Hughes).

Initial systems took advantage of its mapping features to display leaks as they were found by the units and CAM units that had chronic background noise. With GIS becoming increasingly integrated into distribution system operations, the leak data can be used to characterize leaks of specific pipe for future reference, allowing the system to learn. If a system is highly reliable in finding leaks (and not having many false positives) it would appear highly feasible to link the alert about an acoustic detection to a leak repair report that in turn feeds into a main prioritization programme to optimize the selection of mains for renewal.

An opportunity exists to integrate the CAM system with district metering areas (DMAs). The DMAs may already exist or be created on a permanent or temporary basis. In either case, the DMA would serve to help confirm how effectively the CAM system has detected all leaks at any one time. DMAs also provided a method to approximate the loss due to leakage by checking on night flow. By checking the DMA master meter against the aggregate of customer meters (usually through AMR or AMI), total NRW can be assessed and help to quantify apparent water loss.

Chapter 9.6

Establishing the first validated dataset of North American water utility water audit data

A. Chastain-Howley, G. Kunkel and W. Jernigan

The routine compilation of standardized water audits by water utilities is a relatively new practice in North America. The IWA water audit methodology was approved as standard practice through the AWWA in 2001. As the education of utilities has improved and other tools have been made available (such as the AWWA Free Water Audit Software ©) validation has improved. However, with more than 50,000 utilities it is a long and complex process.

The major emphases of this paper are (1) the validity of the data reported by the water utilities and (2) the range of operational and financial performance indicators calculated from the input data. Data Validity is a measure of the accuracy of the audit. There are many terms that may be interchanged, including data confidence, integrity, correctness, accuracy, quality and reliability. All of these terms are synonymous with Data Validity. AWWA developed and published a means for quantifying Data Validity as part of the AWWA Free Water Audit Software © (Chastain-Howley, 2010). Each data input is assigned a grading value of 1–10, based on how a utility's policies and practices match up to a set of grading criteria for a particular data input. This is not meant to be as detailed as a statistical or component analysis. It is a first analysis of validation to discover poor validation and data errors.

The Water Audit Data Validity Score (DV) is calculated based upon the grading of all of the entered components and this value is displayed near the bottom of the Reporting Worksheet of the software. For data inputs that do not apply, a grading of 'n/a' is assigned, and the data input is removed from the calculation of the DV. The DV is then rebased such that the maximum possible score is always 100.

In the context of the water utility's water supply and customer billing operations, certain quantities have greater impact on the water audit than others. Water Supplied (Volume from Own Sources, imported and exported water), Billed Metered Consumption, and Customer Metering Inaccuracies are significant inputs, as any degree of error in these three inputs will more heavily skew the accuracy of the Water Audit results versus an equal degree of error in other data inputs. As a result, the most effective efforts for improvement of Data Validity often involve one or more of these three inputs. The software includes a mechanism to recognize the importance of the above components.

The importance of Data Validity is that water audit data provide the basis upon which water utility managers and governing boards make decisions for investment or deferment of resources for the management of nonrevenue water. Accuracy of information promotes effectiveness in water loss management and revenue recovery in utility systems. Inaccuracy of information promotes misalignment of resources and utility system inefficiencies.

9.6.1 METHODOLOGY

The WLCC Water Audit Software Subcommittee (WASS) requested Water Audit (audit) data from 26 utilities in the United States and Canada. Twenty-three utilities replied with data, and the WASS conducted validation interviews within a scheduled timeline for 21 of those utilities. Data was provided by the utilities for a recent fiscal year (2009 or newer) in the latest version of the software (Version 4.2, June 2010). Each utility also completed a Water System Practices Survey which provided system background information. A matrix of reviewers was developed to assign two committee members to each audit. A validation checklist was developed to guide review of the audits (Figure 9.6.1).

II. Water Supplied

a) Volume from own sources

2. List the number of water source pipelines supplying water to the system (pipelines that convey water from a river, lake, stream, well-field or other source)	
3. List the number of the water source pipelines that are metered?	
4. What is the typical frequency that the source meters are verified? **This information is provided in the Water Utility Practices Survey**. (Remember: meter verification is more than simply calibrating the meter instrumentation and includes steps to confirm the accuracy of the meter's flow measuring capability).	
5. How many meters were found to be with inaccuracy greater than +/- 3% during the past year?	

Figure 9.6.1 Example from Validation Checklist. (*Source*: Andrew Chastain Howley).

Committee reviewers first evaluated the data inputs and their corresponding gradings, looking for abnormalities or inconsistencies in the data. Telephone interviews were then conducted, ranging from 1 to 2 hours, with one or more representatives from each utility to further discuss and scrutinize the data submitted. The primary focus of each telephone interview was evaluation of data sources and grading values, with review of data inputs to determine consistency of reporting. Interviewers questioned the specific policies and practices of the water utility in order to gain a fuller understanding of how data are collected and what quality control measures are in place. Any resulting amendments to data values or their gradings were documented and incorporated into a revised water audit. Once this process was completed for a given audit, the audit was considered 'validated'. Pre- and post-validation audits were physically assembled for comparison and analysis via a database compiler which was also developed by the WASS for the purpose of facilitating water audit data management. A map of locations of utilities with validated water audits is provided below (Figure 9.6.2).

Of the 26 audits submitted, 21 had been validated at the time of publication of this article.

Key performance indicators (KPIs) were statistically reviewed and trends analysed. The basis for trend analysis included the following:

- Total system size – KPIs were compared between systems with greater than and less than 50,000 connections.
- Temperature – KPIs were compared between systems with greater than and less than an average annual temperature of 50°F or 10°C.

- Rainfall – KPIs were compared between systems with greater than and less than an average annual rainfall of 30 inches.
- Number of connections – KPIs were compared between systems with greater than and less than a connection density of 60 connections/mile of main (40 connections/km).

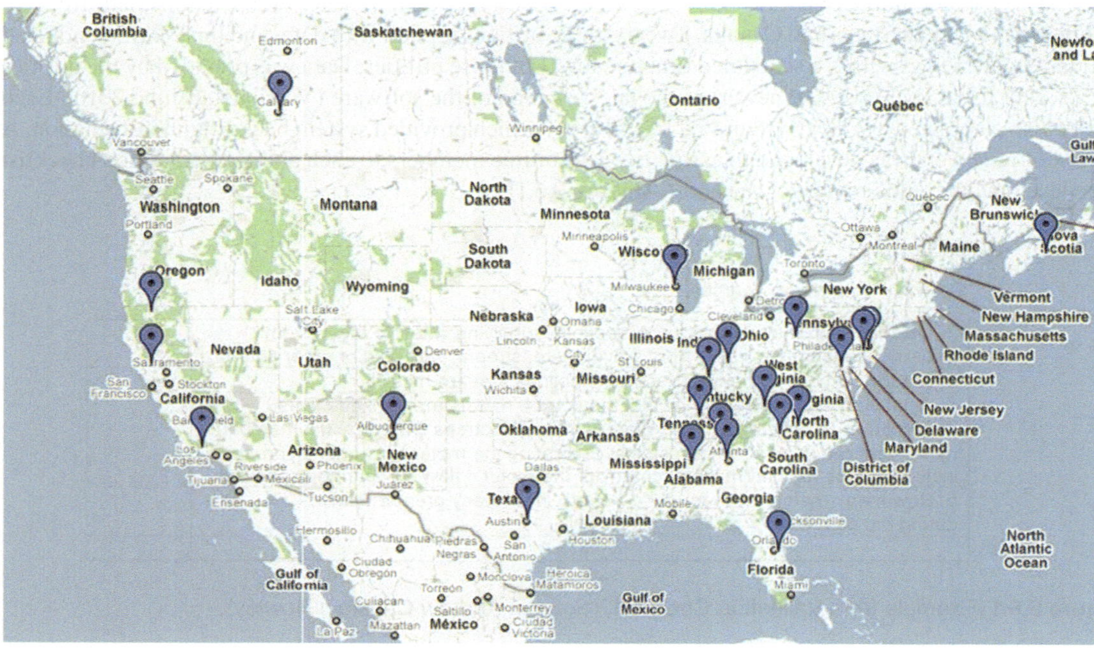

Figure 9.6.2 Map of validated audit locations (*Source*: Andrew Chastain Howley).

9.6.2 RESULTS

The data validity scores for the water utilities in the validated dataset ranged from 52 to 90, and are shown in Figure 9.6.3. In addition to the assessment of the composite scores for the water utilities, gradings for individual components can be compared.

Utilities who import all water do not enter an input for 'Volume from Own Sources' and these appear as a '0' in Figure 9.6.3. There is an average of 7.44 (see Table 9.6.1) for the utilities surveyed. This outlines that most of the utilities conduct electronic calibration of their source meters, but do not conduct any additional flow testing on these meters to prove the flow data.

Table 9.6.2 presents the calculated Key Performance Indicators and Cost Data from the overall validated dataset.

The following Tables (9.6.3 through 9.6.5) outline some basic variations and data trends. The number of connections, average temperature, and connection density are analysed.

A comparison of Key Performance Indicators and Cost Data based on system connection density is presented in Table 9.6.5.

Twenty-one of the 26 submitted audits were validated at the time of publication of this paper. The remaining audits will be incorporated into the next phase of data initiative, which will be in 2012. It is the intention of the WLCC to update the audits currently in the database on an annual basis, as well as to add new utilities to the database each year.

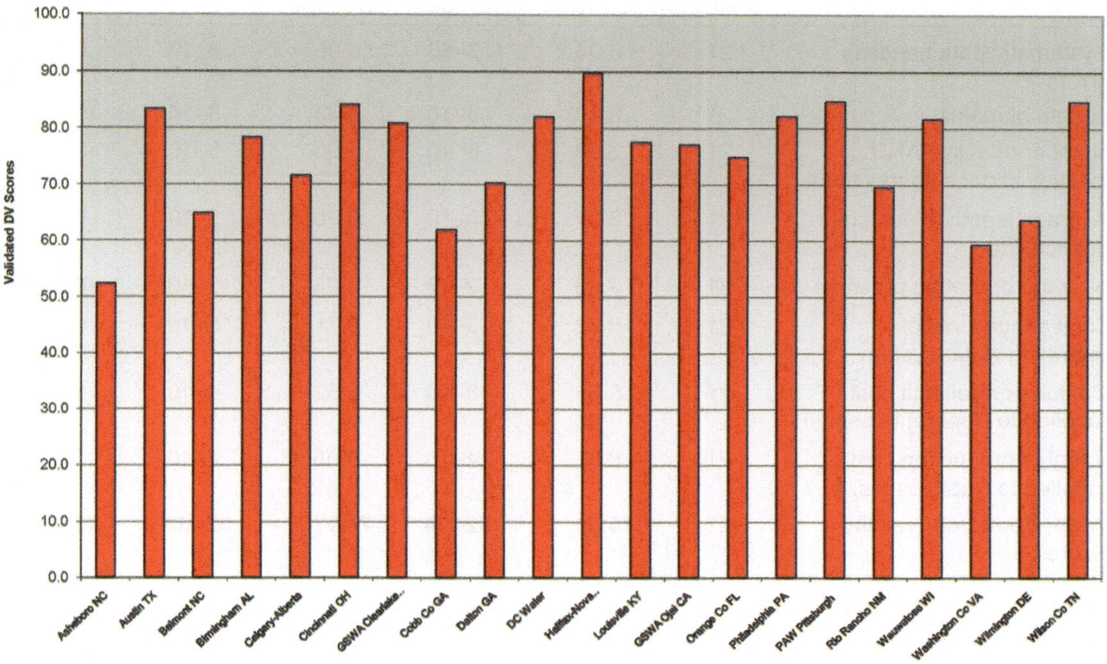

Figure 9.6.3 Validated scores (*Source*: Andrew Chastain Howley).

Table 9.6.1 Pre- vs. post-validation data input statistics.

Data input	Pre-Validation			Post-Validation		Change
	Utilities	Average	Range	Average	Range	
Volume from own sources	16	8.39	2–10	7.44	2–10	(0.95)
Master meter error adjustment	16	6.33	1–10	5.81	2–10	(0.52)
Water imported	12	8.33	2–10	7.75	2–10	(0.58)
Water exported	10	8.85	7–10	8.60	6–10	(0.25)
Billed metered	21	8.35	3–10	8.24	4–10	(0.11)
Billed unmetered	7	8.91	3–10	9.29	6–10	0.38
Unbilled metered	16	7.94	1–10	7.88	1–10	(0.07)
Unbilled unmetered	21	6.58	3–10	6.24	3–9	(0.34)
Unauthorized consumption	21	5.58	5–8	5.43	5–8	(0.15)
Customer metering inaccuracies	21	7.08	3–10	6.81	3–10	(0.27)

(Continued)

Table 9.6.1 Pre- vs. post-validation data input statistics (*Continued*).

Data input	Pre-Validation			Post-Validation		Change
	Utilities	Average	Range	Average	Range	
Systematic data handling errors	21	6.00	3–10	6.05	3–10	0.05
Length of mains	21	8.38	3–10	8.33	3–10	(0.05)
Number of active AND inactive service connections	21	8.38	5–10	7.95	5–10	(0.43)
Average length of customer service line	21	8.15	3–10	9.00	5–10	0.85
Average operating pressure	21	7.38	2–10	7.05	2–10	(0.34)
Total annual cost of operating water system	21	8.62	3–10	8.71	5–10	0.10
Customer retail unit cost (applied to Apparent Losses)	21	8.69	6–10	8.52	6–10	(0.17)
Variable production cost (applied to Real Losses)	21	8.31	4–10	8.05	4–10	(0.26)
Water Audit Data Validity Score	**21**	**78.41**	**52–94**	**74.97**	**52–90**	**(3.44)**

Source: Andrew Chastain Howley.

Table 9.6.2 Calculated Key Performance Indicators – Overall.

Key Performance Indicator	#	Average	Range
NRW as a % by volume	21	22.6%	6.8%–45.5%
NRW as a % by cost	21	10.0%	1.7%–23.0%
NRW–Total annual cost (Million $)	21	$5.81M	$0.04M–$42.97M
Apparent losses (litres/conn/day)	21	***57***	***9–249***
Real losses (litres/conn/day)	18	***240***	***65–567***
Real losses (litres/km of main/day)	3	***4283***	***1518–8159***
Infrastructure Leakage Index (ILI)	21	3.57	1.15–12.68
Water audit data validity score	21	74.97	52.28–89.72

(*Continued*)

Table 9.6.2 Calculated Key Performance Indicators – Overall (*Continued*).

Cost data	#	Average	Range
Annual operating cost (Million $)	21	$51.22M	$1.36M–$224.43M
Annual operating cost [$ per m^3]	21	***$0.88***	***$0.30–$2.15***
Customer retail unit cost [$ per m^3]	21	***$1.21***	***$0.29–$2.21***
Variable production (or import) cost [$ per m^3]	21	***$0.19***	***$0.05–$0.57***

Source: Andrew Chastain Howley.

Table 9.6.3 Comparison of Key Performance Indicators among systems with greater than and less than 50,000 connections.

Key Performance Indicator	# connections <50,000			# connections >50,000		
	#	Average	Range	#	Average	Range
NRW as a % by Volume	10	24.1%	12.2%–45.5%	11	21.4%	6.8%–39.6%
NRW as a % by Cost	10	9.3%	3.1%–17.5%	11	10.6%	1.7%–23.0%
Apparent losses (litres/conn/day)	10	***39***	***9–78***	11	***72***	***24–249***
Real losses (litres/conn/day)	7	**222**	–	9	**250**	**30–567**
Real losses (litres/km of main/day)	10	***4283***	***1518–8159***	11		–
Infrastructure Leakage Index (ILI)	10	3.51	1.24–12.68	11	3.62	1.15–9.89
Water audit data validity score	10	70.44	52.28–84.79	11	79.08	61.92–89.72
Cost data	**#**	**Average**	**Range**	**#**	**Average**	**Range**
Annual operating cost (Million $)	10	9.16	1.36–29.08	11	89.45	24.77–224.43
Annual operating cost [$ per m^3]	10	***1.12***	***0.49–2.15***	11	***0.65***	***0.30–1.15***
Customer retail unit cost [$ per m^3]	10	***1.34***	***0.84–2.21***	11	***1.08***	***0.29–2.08***
Variable production (or import) cost [$ per m^3]	10	***0.26***	***0.09–0.57***	11	***0.13***	***0.05–0.47***

Source: Andrew Chastain Howley.

Table 9.6.4 Comparison of Key Performance Indicators among systems with greater than and less than average annual temperature of 50°F or 10°C.

Key Performance Indicator	Average annual temp >50°F/10°C			Average annual temp <50°F/10°C		
	#	Average	Range	#	Average	Range
NRW as a % by Volume	14	20.9%	6.8%–45.5%	7	26.2%	14.4%–42.9%
NRW as a % by Cost	14	10.7%	3.1%–23.0%	7	8.7%	1.7%–19.1%
Apparent losses (litres/conn/day)	14	***60***	***10–249***	7	***50***	***24–116***
Real losses (litres/conn/day)	11	***203***	***65–470***	7	***297***	***120–567***
Real losses (litres/km of main/day)	3	***4283***	***1518–8159***	0		–
Infrastructure Leakage Index (ILI)	14	2.75	1.15–7.54	7	5.21	2.24–12.68
Water audit data validity score	14	72.62	52.28–84.79	7	79.66	63.79–89.72
Cost data	**#**	**Average**	**Range**	**#**	**Average**	**Range**
Annual operating cost (Million $)	14	38.77	1.36–168.25	7	76.12	5.88–224.43
Annual operating cost [$ per m³]	14	***0.92***	***0.30–2.15***	7	***0.78***	***0.47–1.27***
Customer retail unit cost [$ per m³]	14	***1.26***	***0.72–2.21***	7	***1.10***	***0.29–2.08***
Variable production (or import) cost [$ per m³]	14	***0.30***	***0.05–0.57***	7	***0.12***	***0.05–0.33***

Source: Andrew Chastain Howley.

Table 9.6.5 Comparison of Key Performance Indicators among systems with greater than and less than connection density of 40 connections/kilometre.

Key Performance Indicator	Connection density <40/km			Connection density >40/km		
	#	Average	Range	#	Average	Range
NRW as a % by Volume	9	20.9%	6.8%–45.5%	12	24.0%	12.5%–42.9%
NRW as a % by Cost	9	11.0%	3.2%–17.5%	12	9.3%	1.7%–23.0%
Apparent losses (litres/conn/day)	9	***43***	***10–88***	12	***67***	***9–249***
Real losses (litres/conn/day)	6	***190***	***65–470***	12	***265***	***113–567***
Real losses (litres/km of main/day)	3	***4283***	***1518–8159***	0		–

(*Continued*)

Table 9.6.5 Comparison of Key Performance Indicators among systems with greater than and less than connection density of 40 connections/kilometre (*Continued*).

Key Performance Indicator	Connection density <40/km			Connection density >40/km		
	#	Average	Range	#	Average	Range
Infrastructure Leakage Index (ILI)	9	2.28	1.15–4.27	12	4.53	1.70–12.68
Water Audit Data Validity Score	9	69.98	52.28–84.79	12	78.71	63.79–89.72
Cost data	**#**	**Average**	**Range**	**#**	**Average**	**Range**
Annual operating cost (Million $)	9	43.82	1.36–168.25	12	56.77	1.38–224.43
Annual operating cost [$ per m^3]	9	***0.87***	***0.30–2.15***	12	***0.88***	***0.35–2.07***
Customer retail unit cost [$ per m^3]	9	***1.41***	***1.01–2.21***	12	***1.05***	***0.29–2.08***
Variable production (or import) cost [$ per m^3]	9	***0.26***	***0.06–0.57***	12	***0.23***	***0.05–0.57***

Source: Andrew Chastain Howley.

9.6.2.1 Analysis of DV scores in dataset

The DV scores for the initial dataset largely fall into Levels III and IV of the water loss control planning guide within the AWWA Free Water Audit Software ©; utilities with a DV score in the range of Level III and above represent those with at least basic data collection policies and procedures in place, and have sufficient validity to begin short- and long-term loss control efforts, set long-term reduction targets and utilize the relevant performance indicators.

Average gradings for 'Volume from Own Sources', 'Billed Metered Consumption', and 'Customer Metering Inaccuracies' for the initial dataset were strong. This may stem from the fact that the utilities willing to participate in the initial phase of the project were inherently early adopters of the Water Audit method, and more likely to have already made headway on these three most important aspects of water utility operations. Those utilities in the dataset with lower gradings in these three categories, such as Asheboro and Belmont, reported during the audit interviews about specific improvements to finished-water metering and testing that will result in a significant increase in their grading for 'Volume from Own Sources', and accordingly the DV score in the coming year's audit, which will be included in the next phase of the WLCC data initiative.

Average gradings for 'Customer Metering Inaccuracies' were slightly below those of 'Volume from Own Sources' and 'Billed Metered Consumption', with over one-third of the participating utilities at a score of 5 or less for this category. This may be a reflection on the variability of meter testing and replacement programmes among utilities in general. Some of the utilities in the dataset had regular testing or replacement programmes, but few had both.

The greatest positive and negative adjustments to gradings were observed for 'Volume from Own Sources' (–0.95 points on average) and 'Average Length of Service Line' (+0.85 points on average). For 'Volume from Own Sources', the driving cause for utilities to overstate Data Validity for this input was a general misconception that accuracy testing for finished water meters need only test electronic registration. In many of the utilities, flow verification is not performed in conjunction with testing electronic registration. Both flow and electronics testing must be conducted in order to achieve the highest degree of confidence in finished water meter output, and therefore the highest grading for the same.

For 'Average Length of Service Line', the driving cause for utilities to under-grade this input was a general misconception of the basis for how this distance is measured. For systems who locate customer meters inside the customer's building line (typically for colder climates), this distance is calculated from curb stop to meter. For the majority of the systems in the initial dataset, however, customer meters are typically located at the customer's property line. The distance is therefore '0', and a grading of 10 is assigned by the software.

The averages of the utility DV scores saw a –3.44 point decrease after validation. The fact that there was any change indicates that general understanding of the Data Validity scoring process is still requiring education, even among those utilities who are early-adopters.

9.6.2.2 Trend review

Comparisons can be made among systems in the dataset, on the basis of certain operational and environmental conditions. These comparisons are discussed below, with the caveat that the initial validated dataset is comprised of only 21 utilities, and future expansion of this dataset will more soundly reveal trends among different utilities, based on different factors.

9.6.2.3 System size

On the basis of system size, smaller systems (those with <50,000 connections) showed a slightly higher non-revenue water (NRW) as a percentage by volume (24.1%) as compared to larger systems (21.4%). This is to be expected given smaller systems will have lower system inputs than larger systems. Smaller systems, however, showed a lower actual volume of loss, both real and apparent, on a normalized basis (gal/connection/day). This shows why percentages should not be used to measure water losses. The larger utilities generally have higher use per connection and so any loss percentage appears to be lower than a smaller system. The gallons per connection indicator appears to be much more reliable as a benchmark indicator. Interestingly, the average ILI for smaller and larger systems were very similar, or 3.51 and 3.62, respectively. Since this indicator is determined by the internal dynamics of the individual systems (connections, pressure and miles of main) it is not as susceptible to changes in usage characteristics. It can therefore be used across all systems (although it is not generally used for systems under 3000 connections). Data Validity scores for smaller systems averaged about 70, as compared to about 79 for larger systems, which may reflect the trend of advanced data collection and management systems in larger utilities.

9.6.2.4 Climate

On the basis of system climate, specifically temperature, warmer climates (those with average annual temperature of greater than 50°F or 10°C) showed a lower real loss per connection (203 litres/connection/day) as compared to colder climates (297 litres/connection/day). Apparent Losses were slightly higher in the warmer utilities (60 to 50 litres/connection/day). Likewise, the ILI for colder climates (5.21) was almost double that of warmer climates (2.75). This may be a reflection of harsher ambient ground conditions in colder climates and the propensity for system breaks and leaks. Also, utilities in warmer climates generally put greater emphasis on water conservation, which often leads to more proactive leak detection and water loss reduction programmes.

On the basis of rainfall (table not shown), in drier climates (those with total annual rainfall of greater than 750 mm, or 30 inches) apparent losses were about half the amount and Real Losses were about two-thirds of the amount in drier climates as compared to wetter climates. All these data differences are probably due to the propensity of water conservation efforts in the drier climates and the need for water loss reductions to balance out the fact that these utilities are also asking their customers to reduce water usage. ILI and Data Validity scores were comparable between these two climates.

9.6.2.5 Connection density

On the basis of connection density, less dense systems – those with connection density of less than 40 connections per kilometre, (60 connections per mile) showed slightly less real loss per connection (190 litres/connection/day) as compared to more dense systems (265 litres/connection/day). Since one of the main locations for real water losses is at the service connection, this appears logical. Normalized Apparent Losses (litres/connection/day) were about 57% higher and normalized Real Losses were about 40% higher in more dense systems, as compared to less dense systems. Less dense systems showed an ILI of about half (2.28) that of more dense systems (4.53). Data Validity score was about 12% higher in more dense systems.

9.6.2.6 Cost data

A review of cost data (table not shown) reveals, as expected, a notably higher average variable production cost $*0.34 per m*^{3} ($1.29/1000 gallons) among systems who purchase (import) 100% of water supplied, versus those who produce some or all of their water supplied *$0.15 per m*^{3} ($0.55/1000 gallons). Average customer retail unit cost between these two groups was comparable *$1.17 to $1.30/m*^{3} ($4.46 and $4.92/1000 gallons).

9.6.2.7 Improving Data Validity

Improving Data Validity comes from a combination of top-down (records analysis and calculations) and bottom-up (field measurement) efforts. Ultimately, the reliability of the top-down Water Audit is improved by incrementally incorporating bottom-up approaches to field-verify assumptions and estimations (Thornton *et al.*, 2009). As mentioned above, certain components exert a stronger effect than others in the water audit. Initial bottom-up efforts for improving Data Validity should be focused on these significant components. For 'Volume from Own Sources', focus should first be that all finished-water inputs to the distribution system are metered with meter readings digitally archived, and second that those meters are tested for accuracy of both flow measurement and electronic registration at least annually. Data should be reviewed regularly and adjusted to account for any data gaps that can occur if instruments are out of service for periods of time. For 'Billed Metered Consumption' and 'Customer Metering Inaccuracies', focus should be on the minimization of estimated billings, utilization of billing software that can be electronically queried for meter data, and the development of a routine testing programme that dictates a meter replacement protocol based on cumulative consumption and meter age.

9.6.3 CONCLUSIONS

Ranges and averages for Data Validity as presented in this paper can be utilized for reference. However, this is an initial dataset intended for annual updating. It is also expected that the initial dataset will be expanded with additional participating utilities. At least three years of data compilation and analysis will be needed to represent a robust dataset for stronger benchmarking. More utilities will be invited to participate in future phases, but only to the extent that the reported utility data can be validated.

Data Validity scores are generally strong in this initial dataset, but the dataset represents early-adopters so the effect of expanding the dataset on the average Data Validity Score may be difficult to predict.

REFERENCES

Chastain-Howley A., Goff D., Kunkel G., Lalonde A. and Sayers, D. (2010). AWWA Free Water Audit Software©, version 4.2.

Thornton, Julian *et al.* (2009). Manual of Water Supply Practices – M36, Water Audits and Loss Control Programs, 3rd edn. American Water Works Association, Denver, CO.

Jernigan W., Chastain-Howley A., Cobb M., Goff D., Kunkel G., Lalonde A. and Sayers D. (2011). Assembling validated water audit data for reliable utility benchmarking. *AWWA Distribution System Symposium*, Nashville, TN.

IWA Publishing's authorised EU representative for General Product Safety Regulations is Diane D'Arras, 15 rue Duret, 75116 Paris, France, e-mail: safety@iwap.co.uk.

Printed and bound by CPI Group (UK) Ltd, Croydon, CR0 4YY
01/04/2026
02082208-0002